MALDON
7116

2 3 MAR 2017

Please return this book on or before the date shown above. To renew go to www.essex.gov.uk/libraries, ring 0345 603 7628 or go to any Essex library.

GW00503706

Frontispiece: Thomas Newcomen's Engine erected at Dudley Castle in 1712 and modelled by the author.

THE EARLY DEVELOPMENT OF THE STEAM ENGINE

by

DAVID K. HULSE

Published by TEE Publishing, The Fosse,
Fosse Way, Leamington Spa CV31 1XN
Telephone 01926 614101

ISBN 1 85761 107 1

ACKNOWLEDGEMENTS

While writing this acknowledgement of all the help I have received after twenty-four years of research undertaken to construct my model reproductions of eighteenth century engines, comes the realisation of just how many people I have met or corresponded with since this project began in 1974. The work has brought about many friendships and acquaintanceships with people who have not spared themselves to provide information which has been brought together as a chronicle into the development of the steam engine from 1700 until c1800. Not only have I received documentary help but many companies have gone out of their way to supply that extra special material which was required to make these model engines.

In an attempt not to miss thanking all those who have helped, I have decided to list each engine and acknowledge those who have been involved in each engine's production.

Thomas Newcomen's Dudley Castle Engine – John Allen, the chairman of the Midlands Branch of the Newcomen Society, has researched Newcomen and his family for many years. He has also been responsible for the Newcomen Engine which has been built at the Black Country Museum. John shared this research with me, greatly helping to recreate the engine in a miniature form; and Eric Griffith, the design director of Royal Doulton, for sculpturing the ceramic figure dressed in early eighteenth century costume on this engine.

The Smethwick Engine – Jim Andrew, of the Birmingham Museum of Science and Industry, who has researched this engine for more than ten years. This research was submitted to the University of Birmingham where he was awarded his Ph.D. It was this extensive research which enabled me to construct a faithful representation of how the engine was built by James Watt in 1779; Nicholas Cresswell, of Glebe Engineering, for making the welded steel frames for this engine and also for supplying the steel blanks for the cylinder and the pump barrels, and also for the laser profiling of the small chain links; Martin Tidmarsh, of Engineering Services, for the electrical control to the engines; and to Steve Leadly, a sculptor from Royal Doulton, for the ceramic figures on four of the model engines.

Grateful thanks must also be given to William Tranter, associate editor of *Engineering in Miniature,* for his very patient editing of my text, and to Chris Deith, Managing Editor of TEE Publishing, for serializing this work in *Engineering in Miniature* and for his decision to bring all the research together into two volumes.

Finally, a big thank you to someone who, for twenty-four years, has allowed me the freedom to pursue this research to a conclusion, first checking, and many times rewriting, my text into a more readable form – my dear wife Julie.

David K Hulse
January 1999

PREFACE

When it was decided that David Hulse's series, *The Early Development of the Steam Engine*, would be published in *Engineering in Miniature*, we had little idea of just how monumental and popular a work it would become.

It all began as a discussion about David's superb models of the engines and engine houses, and how interesting it would be for our readers if he could write about his construction methods and the tooling required to make the models. Intense interest was shown in his display of the model engines at the Midlands Model Engineering Exhibition with people actually queueing to talk to David about his unique brickmaking machine and to see it being operated! From this small beginning has sprung a massive work which has grown larger than anyone could ever have imagined, as David has continued to study and research these fascinating early steam engines which, until now, have never been fully described clearly and completely.

The articles published in *Engineering in Miniature* have become our longest series and, while the author continues to unearth further fascinating glimpses of our industrial heritage, the series has not yet finished. Certainly the number of favourable letters and other correspondence with our readers has proven just how popular and readable his series has become.

The amount of research and study by David is staggering, but his miniature engines, which are not so 'miniature' as you might think with some of the clear Perspex cases protecting the models being nearly two metres high by one metre square, are breathtaking. His dedication to detail and his patience is very evident, with each component re-created exactly as on the prototype engines, right down to every nut and bolt, square ones, of course! The timber work of the engines and their houses is also portrayed in minute detail, with the scrapes and gouges made on the timber by the artisans of the eighteenth century being reproduced faithfully by the author.

The bricklaying of the engine houses took months of patience, carefully placing each brick in position with 'mortar' made from wall tile adhesive. Of course, as David mentions in this book, the development of the machine to produce the bricks and tiles was quite a task also! There is no doubt in anyone's mind that the models are well deserving of their first prizes in the competition classes of the Midlands Model Engineering Exhibition.

This book has arisen in response to the repeated requests by our readers that a book be written expanding even further upon the *Engineering in Miniature* articles, and it is a pleasure that TEE Publishing is associated with David's work.

Chris Deith
Managing Editor
Engineering in Miniature

CONTENTS

LIST OF ILLUSTRATIONS

LIST OF ILLUSTRATIONS

INTRODUCTION

In 1974, after almost twenty years of restoring and riding motorcycles, I decided that I would research and construct in miniature what I considered to be the most important steam engines built between the years 1712 and 1804. These historical engines really did pave the way for the industrial revolution. In chronological order, these engines are:

1) Thomas Newcomen's Dudley Castle Engine - the world's first recorded steam engine built in 1712.

2) James Watt's 'Smethwick Engine', one of the first engines designed by James Watt and used to pump water up a series of canal locks. This engine went into service in May 1779.

3) Another engine by James Watt – this engine was specifically designed to produce rotary motion and has become known as the Lap Engine because of its task of driving machinery used to lap and polish small manufactured articles. The Lap Engine was made in 1788.

4) A rotary beam engine designed to drive a worsted mill at Arnold, Nottingham. The engine was designed by Francis Thompson, an engineer from Ashover, Derbyshire. He took out a patent for this unique engine in 1792 and the engine's unusual features were designed to achieve success without infringing patents held by James Watt on steam engine design. The engine at Arnold was built in 1797.

5) An engine designed by Richard Trevithick and was built in 1804 to provide the rotary power to drive a dye house at Lambeth and is the first high pressure steam engine to work with its powering cylinder positioned horizontally.

6) This is a model of a high pressure beam engine built circa 1860. This miniature engine does not represent any particular evolutionary step – I generally refer to this as my apprentice piece, but it does show how engine power increased and the engine's physical size decreased.

The six miniature engines briefly described here took a little over twenty four years to complete and over this time some 21,000 hours were spent on their construction. As well as the actual model engines, some special tools and machinery had to be designed and made because many of the items needed were not commercially available. It is not my intention to describe how each part of these models was made in detail, but I will describe how some of the more unusual parts which are not generally used in miniature engine construction were made, for example, how over 141,500 scaled ceramic bricks were made and then laid individually – no, this is not a misprint, that is how many bricks were used to complete this task!

The miniature engines were not made in chronological order, because I wanted to learn the craft of modelmaking as I went along, so the models were made in the following order:

1

1) A high pressure beam engine – the apprentice piece.
2) The Lap Engine.
3) Thomas Newcomen's Dudley Castle Engine.
4) The Smethwick Engine.
5) The Richard Trevithick Engine.
6) The Arnold Mill Engine.

On completion of my first engine, which had been constructed without the use of castings, I felt confident that I could make a start on the remaining miniature engines.

The five engines which I chose to construct in miniature were considered to be very important steps in the advancement of the industrial revolution, so from the start I decided that these engines must be made with great accuracy and with minute attention to detail. With this thought in mind, no component part was going to be made of any materials other than that which would have been used in the year of the engines' manufacture. These eighteenth century engines were always a structural part of the actual building intended to house them, so it was decided that these buildings also should be constructed using miniature red clay ceramic bricks. All the component parts of the five model engines were hand made, including the nuts, pins and washers. Also the woodwork of these engines has been either cut to simulate pit sawing or, on the larger parts, hewn to its correct size and shape to give the component the appearance of a piece of wood made by a craftsman wielding an adze.

Constructing the miniature engines in this way does have one very big advantage: when it is complete the visual appearance does not have to be finished by modern paints or plastics. The authentic materials do give the appearance of having been made some two hundred years ago.

James Watt's Lap Engine was the first model to be completed. This was after almost nine months of research and detailed planning. The miniature bricks for this engine were made using a very tried and tested clay forming process. All of the metal parts were made by hand. On none of the model engines are castings used; all the component parts which were originally cast have been fabricated and lightly grit blasted to give the appearance of iron castings. The method used to produce the bricks for the Lap Engine proved to be very time consuming and tedious and, after eleven thousand bricks, I thought the remaining engines, which needed so many more bricks, was simply 'not on'. I would have to have a rethink of how they were to be produced.

In my day to day job as Royal Doulton's Chief Development Engineer, I was privileged to have access to all the latest manufacturing techniques and in 1983 a new manufacturing method was being developed. This method is now widely used and could, without doubt, be classed as the industrial revolution of the pottery industry.

The pottery, such as dinner plates, side plates and saucers, now made at Royal Doulton, is now produced by pressing dry powdered clay between steel and plastic dies. This process is done at a very high pressure – a dinner plate of about 10" diameter requires a force of 500 tons.

With the advent of this new technology, I began to wonder if my small bricks could be made by this method. I spent many hours at my drawing board designing a miniature version of these enormous machines complete with the miniature brick dies.

With the design complete, I set about making this machine. I was very

2

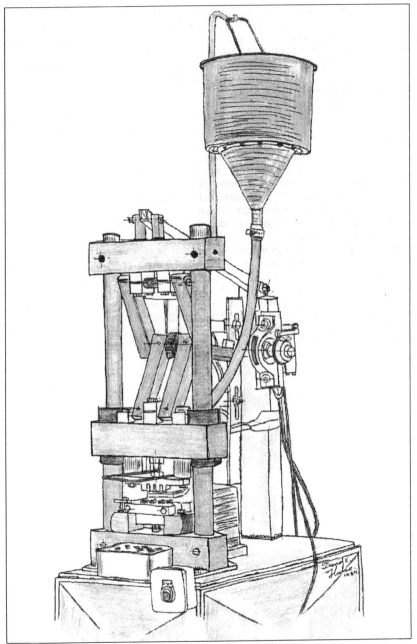

Fig. 1: The miniature brickmaking machine designed and built by the author. This machine was first demonstrated at the 1995 Midlands Model Engineering Exhibition.

eager to try this new innovation in clay forming and, to my utter amazement, it produced bricks of good quality. The first filling of the dies with dry powdered clay was excellent. This is now my standard method of brick making and has been used on all the remaining miniature engine houses.

I doubt whether the task I set myself of producing the model engines could have been achieved without this new technology. I would like to state that even to form my miniature bricks in a four cavity die, the press has to exert a force of 5.5 tons. The machine which was designed and made to produce the miniature bricks is shown in Fig. 1. I do not intend to go into the details of the machine at this stage – it will be described in full later, together with how the process works.

The first four miniatures (in chronological order) are of engines which were powered by a vacuum. This vacuum was created by the condensation of steam. The engines were all known as 'atmospheric engines'. The following chapter will explain why a vacuum was used and not the more obvious property of the expansive force of high temperature steam.

So much information has now been amassed that I decided to re-write the description published in 1985 in *Engineering in Miniature* and this was published monthly in *Engineering in Miniature* over a period of six or so years. As we progress through all this research and the description of construction of the model engines, I will explain how the more interesting parts were made and how many of the problems were overcome. My writings will describe the experiments and scientific knowledge which led to Thomas Newcomen developing the world's first steam engine in 1712. Then I will try to describe what happened throughout the eighteenth century, culminating in Richard Trevithick's high pressure horizontal steam engine of 1804.

David K. Hulse
1999

CHAPTER 1

THE ANCESTRY

SOME EARLY ATTEMPTS TO HARNESS STEAM POWER

The steam engine can be claimed without doubt as the greatest single engineering invention of the industrial revolution. It certainly heralded its start, and the credit for this magnificent invention should go to Thomas Newcomen of Dartmouth, Devon who was born in 1663/4 and who lived until 1729. In 1712, Thomas Newcomen developed and erected an engine to pump water from a coal mine at Tipton in South Staffordshire and this engine has become widely known as the Dudley Castle Engine.

But who invented the steam engine? Many people give the answer as James Watt. We are told that he discovered the force of steam whilst observing a boiling kettle. This is quite wrong, as James Watt only made many improvements to the steam engine. His greatest single improvement was the separate condenser; this he patented in 1769, almost sixty years after Thomas Newcomen set his first engine to work at Dudley Castle.

The Newcomen Engine worked by condensing steam to create a vacuum. It was this vacuum which pulled the piston along the bore of a large brass cylinder. It was the first of the piston and cylinder type engines which worked because of the pressure of the earth's atmosphere. The only known record of this great discovery is an engraving drawn in 1719 by a Wolverhampton file maker, Thomas Barney. Thomas Newcomen in the early eighteenth century did not receive much public credit for his invention.

An eminent Swedish engineer, Martin Triewald, who visited England about this time, met and worked alongside Newcomen, stated in 1734:

'For ten consecutive years Mr Newcomen worked at this fire-engine which never would have exhibited the desired effect, unless Almighty God had caused a lucky incident to take place. It happened at the last attempt to make the model work that a more than wished-for effect was suddenly caused by the following strange event. The cold water, which was allowed to flow into a lead-case embracing the cylinder, pierced through an imperfection which had been mended with tin-solder. The heat of the steam caused the tin-solder to melt and thus opened a way for the cold water, which rushed into the cylinder and immediately condensed the steam, creating such a vacuum that the weight, attached to the little beam, which was supposed to represent the weight of the water in the pumps, proved to be so insufficient that the air, which pressed with a tremendous power on the piston, caused its chain to break and the piston to crush the bottom of the cylinder as well as the lid of the small boiler. The hot water which flowed everywhere thus convincing even the very senses of the onlookers that they had discovered an incomparable powerful force which had hitherto been entirely unknown in nature – at least, no-one had ever suspected that it could originate in this way.'

5

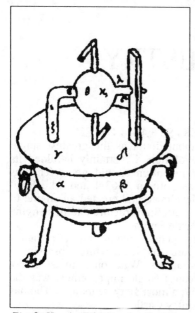

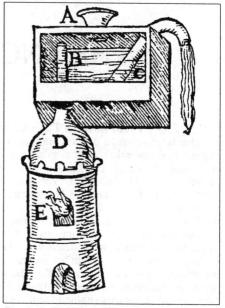

Fig. 2: Hero's 'Whirling Aeolipyle', circa AD100 from the Burnley Ms 81 (sixteenth century) in the British Museum.

Fig. 3: Della Porta's Steam Pressure Apparatus from his Spiritali 1606. The steam generated in the boiler forced water up a pipe.

Hero of Alexandria

Hero of Alexandria, also recorded and known as Heron of Alexandria, was a Greek geometer and writer who lived almost two thousand years ago. The exact date is conjectural some scholars tend to date him as 150BC and others later; the generally accepted date for his steam turbine is 100AD. Hero wrote a mathematical work entitled *The Pneumatica*. In this work, he describes a device which was placed upon a fire to boil. He called this device an 'aeolipyle'. This was the first recorded use of the expansive force of steam. Today we would call the 'aeolipyle' a reaction turbine. It most probably worked by steam which was raised by boiling water in a closed vessel and the steam then entered a hollow sphere by way of a vertical pipe; steam then escaped through the two tangential pipes attached to this sphere and, on escaping, the steam created a reaction which revolved the sphere in a continuous rotary motion. This device became known as Hero's 'Whirling Aeolipyle.'

Steam pressure apparatus

After this first recorded experiment with the power of steam, almost 1,600 years passed before the apparatus shown in Fig. 3 was made. This apparatus is described by Giovanni Battista Della Porta (1538-1615) in a treatise on pneumatics written in 1606. He boiled some water in the wine flask D and the steam which was generated flowed through the pipe B into the closed space above the water in the small tank. The increasing pressure forced the water out along the tube C, the only means of escape.

In another experiment, Della Porta filled a wine flask with steam and then plunged this flask upside down into a bowl of cold water and a partial vacuum was formed which then drew water into the vessel via a tube causing the steam to condense. Porta recorded this phenomenon, but did not appear to understand that was happening. However, it was on this principle which engines were to work a hundred years later.

The Savery and Newcomen engines worked by condensing steam to create a vacuum and the weight of the earth's atmosphere acted upon the vacuum which had been created.

Salomon de Caus

The next recorded experiment on the properties of steam is shown in the sketch of an apparatus by Salomon de Caus, who was an engineer and architect to Louis XIII, King of France. He came to England in 1612 and was employed by the Prince of Wales in the ornamentation of the gardens of his house at Richmond. While at Richmond, he produced several works but the one of any interest in the understanding of the properties of steam was written in 1615 in which he describes the apparatus shown below.

He begins with a definition of the four elements – fire, air, water and earth. He then goes on to say that the sphere, shown in the sketch, was made of copper. The sphere he described was believed to be one or two feet in diameter.

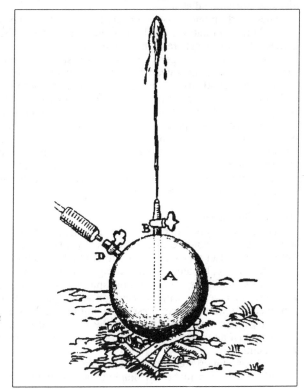

Fig. 4: De Caus' Steam Pressure Ball and Fountain made in c1611, from his 'Les Raisons de Forces Mouvantes', 1615.

This vessel was then filled up to a third full of water before being placed over a fire to boil, usually for between three to four minutes.

De Caus did several experiments with this apparatus and each time his results gave him a better understanding of the pressure of steam and the evaporation of the boiling water.

The surprising theme amongst these early philosophers was that none of them seemed to visualize that there could be any application of their experiments outside the laboratory. They all ceased experimentation before their ideas could be used in a full scale trial. These experimental results appeared enough to satisfy the intellectual understanding of the time.

David Ramsey

The power of steam did not seem to have any practical use until 1630 when a patent or especial privilege was granted by Charles I to David Ramsey, who was one of the grooms of the Privy Chamber.

It is said that he invented nine ideas which the patent covered. The two which are of interest to us deal with the evolution of the steam engine. They read as follows: 'raising water from low pitts by fire, moving mills on standing waters by continual motion, without the help of wind, weight or horse.' This patent was dated 21st January 1630 and was granted for a period of fourteen years, but the evidence suggests that David Ramsey never developed any of his ideas on pressurized steam.

These early pioneers were hindered in their development of steam power by the technical limitations of the seventeenth century. They were only able to harness the power generated by pressurized steam on a small scale. It was not possible to produce a large vessel which was strong enough to withstand the high force which steam generated when confined in a large closed volume or space.

By about 1630, all the indications were pointing to the eventual harnessing and practical use of high pressure steam. However, in Italy, experiments were being conducted on suction – lifting water vertically in stand pipes. In 1641 a great discovery took place. This was that the earth's atmosphere had an actual pressure. This pressure was the governing factor in all suction lifting where a vacuum has to be created and it was becoming evident that the earth's atmospheric pressure controlled the degree of vacuum which could be created. The maximum vacuum which could be created would only equal the pressure of the earth's atmosphere.

Italy to Germany

The next recorded experiment to demonstrate that the earth had a constant pressure which could be calculated was by a German engineer by the name of Otto von Guericke (1602-1686). He studied at Leyden University and, after much travelling, became a military engineer and Burgomaster of Magdeburg.

In about 1650, he made a hollow sphere from copper; he then evacuated the air from this vessel with a hand operated pump. When the air had been removed, the atmospheric pressure acting on the outside collapsed this sphere inwards, again demonstrating that the earth had pressure which could then press against an expanse which had the air inside evacuated. We all know this as a vacuum.

In 1654, Guericke conducted another experiment in which he made a cylinder of fifteen inches in diameter and twenty inches high. Into this, he

Fig. 5: Von Guericke's Experiment with Vacuum c. 1672. Twenty men are trying to haul up the piston with a rope attached to it. A vacuum is created in the cylinder holding the piston at the bottom against the efforts of the men.

fitted a close fitting piston; this is more clearly seen by studying the illustration shown which was drawn in 1672.

The cylinder and piston were attached to a strong wooden frame. The man, working by the side of the upright, was operating a pump which evacuated the air from beneath the piston. It is said that twenty men were unable to move the piston along the bore of the cylinder against the vacuum created.

If the vacuum operating in this way could have been reliably re-created, Guericke would have had the prospect of producing an engine, but nearly sixty years had to pass before we were to see a piston moving with periodic motion inside a cylinder. This was to be the Newcomen Engine of 1712.

Robert Boyle (1627 - 1691)

The man who put together the findings of Torricelli, Pascal and Guericke was the Hon. Robert Boyle. After returning from his continental travels, he conducted further experiments both in London and Oxford. He was trying to formulate the laws of gaseous pressure. With his assistant, Robert Hooke (1635-1703), he designed and made an improved type of air pump. This pump is preserved by the Royal Society and is the prototype of those used to this day. With the aid of this pump, he was able to formulate the law of gas pressures and this is of course known as *Boyle's Law*.

These three early scientists clearly showed that the earth could exert a force against a vacuum. But how could this vacuum be reliably reproduced?

One surprising method which was tried was to produce a controlled explosion. A small charge of powder was placed inside a closed vessel; on ignition, the air was expelled through a 'non-return' valve which then produced a vacuum inside this vessel. In 1661. Sir Samuel Morland was granted a patent for an engine to raise water from mines 'by the force of air and power conjointly.' It is not known what the engine which Morland proposed looked like, or even if one was actually made. Many other people tried to create a vacuum by a controlled explosion, but nobody managed to produce the continuous vacuum which was so essential to move the piston inside the cylinder, which could then be capable of powering an engine.

Denis Papin (1647 - c.1712)

Papin had the distinction of working with the great Dutch astronomer, Christian Huygens (1629-1695). He continued with the series of experiments, the aim of which was to try to create a continuous vacuum which could move a piston inside a cylinder. Again, an explosion was tried, but this time the piston was fitted into the bore of a cylinder and a small explosive charge expelled the air through a 'one-way' valve. With the air removed, a vacuum was created. This vacuum enabled the piston to move through one stroke to the other end of the cylinder. But what was needed was a constant vacuum which could be used to make the piston move in a cyclic way – creating a vacuum by explosions was proving to be too difficult. In 1675, to escape religious persecution, Denis Papin (who was a Huguenot) came to London and worked first with Robert Boyle and then with Robert Hooke and, through this association, he did work for the Royal Society.

Papin turned his thoughts to creating a vacuum by condensing steam inside a closed vessel and between 1690 and 1695 he constructed the apparatus shown in the diagram Fig. 6 opposite – this consisted of the thin metal cylinder A and into this he fitted a piston B which was allowed to move freely. A small quantity of water was put into the bottom of this cylinder, the piston was then moved so that it just touched the water, then all the air was expelled through a hole in the piston. With the air removed, the piston could be sealed by the plug M. The whole apparatus was then placed upon a fire to boil. When the water turned to steam, the piston rose to the top of the cylinder and in this raised position, the catch E could be engaged. This then prevented the piston from falling to the original water level.

The diagram has been redrawn in the hope that Denis Papin's important contribution for the search into creating a vacuum can be better understood. This is shown in Fig. 7 opposite.

The whole apparatus was removed from the fire and allowed to cool but, on cooling, a vacuum formed beneath the piston and when the catch E was released, this vacuum drew the piston through a complete stroke. It also lifted some weights which were attached to the cord by the use of two pulleys. Papin

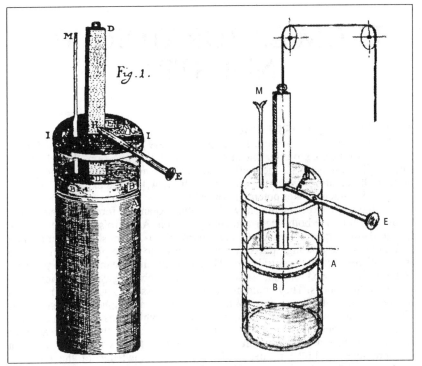

Figs. 6 and 7: Left, Papin's drawing of 1690. Right, Papin's Vacuum Apparatus with a piston in a cylinder, from his 'Acta Eruditorum' 1690, redrawn for clarity.

at last had found a method of creating a vacuum by condensing the steam within a closed vessel – 'the cylinder'.

This apparatus, or 'engine' as it is sometimes referred, was only on a small scale and Papin did not pursue his idea by making this machine on a larger scale. However, he had clearly demonstrated the principle on which the atmospheric engine was eventually to work. It was Thomas Newcomen who had such success with this engine in 1712.

CHAPTER 2

FROM LABORATORY TO INDUSTRY

A SOLUTION ALMOST IN SIGHT

Before proceeding to describe Captain Savery's work and 'The Miner's Friend', I would like to offer one answer to a puzzling question. Why did these early philosophers focus their efforts on harnessing the earth's atmospheric pressure by creating a vacuum when they all knew that the expansive force of steam had a much greater potential power'?

All the early experiments showed that the expansive force of steam could be used to produce both linear and rotary motion. As Hero's aeolipile was the first to demonstrate this principle, these early philosophers showed, on a small scale, that steam at a high pressure could be used to move pistons and rotate a small sphere around a central axis. They thought that steam at high pressure could only be used on this small scale; the technological limitations had governed their ideas. They could not make large vessels to withstand the forces of steam under high pressure. When the earth's atmospheric pressure was discovered, a new branch of research was begun which was to devise a means of creating a vacuum, which was eventually achieved by the condensation of steam. To devise a powering force by this method, large volumes of steam at low pressure would be needed – the technical knowledge of the day was adequate and these low pressure, high volume steam generators could be made. These early pioneers simply copied the stills which were used for the distillation of spirits.

Quite simply, it had been the technological limitations of the era which had been the guiding factor in presenting man with a powering force for machinery which he was able to both use and understand.

Almost 1700 years had passed from the first experiment and it was not until the late 1690s that a machine was designed on a large scale, the practical use of which was to extract water from mine workings and allow miners to work without the constant fear of flooding.

Thomas Savery and his steam pump

A single goal was beginning to be achieved by the 1690s – all the centuries of trial and experimentation were coming together with the production of a machine which could be used to pump water from deep pits such as mine workings. Thomas Savery was the man who combined all the theories explained above and produced a machine big enough to remove the water from the mines. He was a Devonian – the exact year of his birth is not known but it is thought to be 1650.

In July 1698, a patent for Savery's steam pump was granted for England and Wales. It is at this date that we first see Thomas Savery referred to as Captain – where he acquired this title is not known. The most probable source is that the mine superintendents, or bosses, in Cornwall were usually referred

Fig. 8: A sketch by Savery of his experimental engine presented to the Royal Society on 14th June 1699.

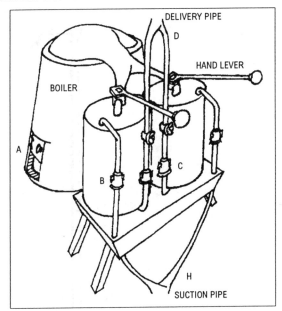

to as mine captains. Perhaps this title was given to him in recognition of his efforts to drain the mine workings of water.

This historic patent granted to Savery in 1698 was for a period of fourteen years and it read 'Raising water by the impellent force of fire.' Thomas Savery only lived for three years after his patent expired and he was buried at St Giles Church, Camberwell, Surrey on 22nd May 1715. He must be remembered for combining all the preceding theories relating to steam power and he was the producer of engines large enough to be used for such practical purposes as mine drainage. The last twenty-five years of the seventeenth century must have been a furious time for the application of this common pursuit.

The Miner's Friend

Thomas Savery wrote a book called *The Miner's Friend*. This was published in 1702 and has a description of an engine for 'raising water by fire, and how to fix it into mines.' This book was published by A. Baldwin in Warwick Lane and it sold at most booksellers for one shilling. The reader is left in no doubt that Captain Savery had a high opinion of himself some of his claims are very exaggerated for example he states that 'my engine is unlimited, and will raise your water five hundred or one thousand feet, were any mine so deep.' In 1702 this claim would have been impossible!

How the Savery pump worked

On 14th June 1699, Savery described to the Royal Society, how his engine worked. The sketch he used is shown above with annotation added by the author. This engine is generally known as the June Engine.

The steam was transferred from the boiler marked A, through the odd shaped pipe work into the containers marked B and C; container B was partly

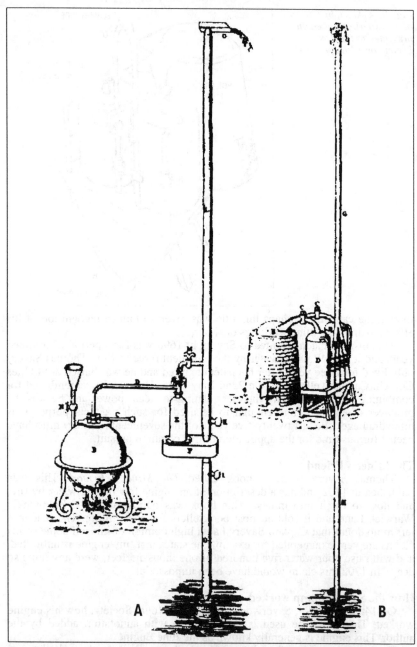

Fig. 9: Early Savery pumps c1698. A – pump with single boiler and receiver.
B – pump with single boiler and two receivers.

filled with water, while container C was empty. The incoming steam pressure expelled the water from the container B along the vertical stand pipe D, and at the same time filled container C with steam. Cold water was then poured over the outside of container C and upon cooling, a vacuum was formed. This then drew water through pipe H to partly fill C; and with the aid of the hand operated levers, was made to alternate between the containers B and C.

The water was expelled vertically from one container and drawn into the other from below the vessel. This was the consummate invention, as Savery had combined all the intellectual thoughts of the past 1700 years into this simple looking sketch. He had taken the great step out of the scientist's laboratory and was trying to put all the theories together to make a machine large enough to be usefully employed in the mines.

To construct his new invention, Savery had to rely upon the craftsmen of the day; but the new methods of manufacture demanded new skills, though the exacting standards laid down by Savery do seem to have been achieved. The cocks and pipework were made of brass, and the receivers and the spherical boilers, sitting upon their brickwork furnaces, were of beaten copper.

One of Savery's first engines which was built on a large scale is shown in Fig. 9.

About 1700, Savery opened what was the world's first steam pump manufactory. This was in Salisbury Court, which is between Fleet Street and the River Thames, but Savery's high hopes did not materialize and the site was abandoned in about 1705. The truth was that to extract water from deep mines, the Savery Steam Pump was not going to be the 'The Miners Friend.'

The earth's atmospheric pressure limited the suction, and the force needed to raise water great distances vertically exceeded the strength by which the boilers could be manufactured.

But his work was on a large scale and could be used for realistic purposes; it was not a small toy in the laboratory unlike his predecessors who only demonstrated principles. His final efforts are shown in Fig. 10 overleaf; clearly he was thinking of very practical uses and not the small scale experiments of his predecessors.

To advance scientifically, the scholars and engineers had to return to the principles which could he understood and controlled. This was to use the earth's atmosphere pressure, acting against a created vacuum, to move a piston within the bore of a cylinder. The man who was to demonstrate this to such great effect, was Thomas Newcomen, by creating in 1712 the world's first commercial steam engine.

THE HISTORIC YEAR OF 1712

The year of 1712 must rank as one of the most historic years in the history of mankind, as it was then that the world's first commercial steam engine started to pump water from a coal mine at Tipton in South Staffordshire. The working of this steam engine can, without doubt, be claimed as the most important single event in the dawning industrial revolution – it enabled miners to extract the coal and mineral wealth from far greater depths than had been possible before.

This was the beginning of industrial expansion in this country and the world at large. The man responsible for this great achievement was Thomas Newcomen (1663/4-1729) of Dartmouth, Devon.

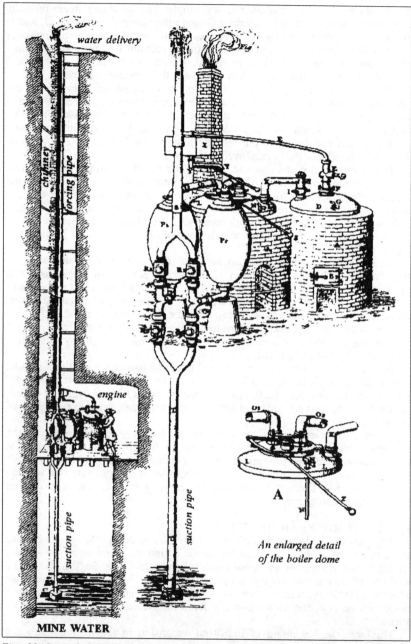

water delivery

chimney

forcing pipe

engine

suction pipe

suction pipe

*An enlarged detail
of the boiler dome*

MINE WATER

Fig. 10: Savery's later design showing twin boilers and receivers. By operating the hand lever Z, a continuous flow of water could be achieved.

In 1712, a steam engine was put to work at a South Staffordshire colliery pumping water from coal mine workings one hundred and fifty feet from the surface. It was the first engine to be built with a piston moving in a vertically positioned cylinder.

The piston was connected to a massive overhead rocking beam by chains. This beam then raised and lowered the pump rods operating the pumps below the engine, which then forced the floodwater to the surface.

This type of engine quickly spread to the many mining areas of Britain and, when news of this invention became more widely known, it also spread to many other countries, notably Europe and Scandinavia.

Newcomen's invention was so practical that it remained unchanged for almost sixty years. Many minor changes were made by notable engineers but never straying far from the basic design that Thomas Newcomen invented in 1712.

This engine was successful because its excellent design took into account the technological limitations of the day; quite simply, it could be made with the tools and skills available in 1712. It worked by creating a vacuum against which the earth's atmospheric pressure could act, and not by the force generated by high pressure steam which was proposed by Newcomen's predecessors.

So technically advanced was his design that the 1712 engine was constructed to operate with automatic valve gear. This can be seen on Thomas Barney's engraving as Fig. 11 on the next page.

It is said that James Watt invented the steam engine by observing a boiling kettle – *this is quite wrong!* James Watt only made detailed improvements to the steam engine sixty years *after* Thomas Newcomen used his first engine at Tipton.

Although Newcomen's work was of great importance, not much is known about him and not even a portrait survives. A learned society was formed in 1920 with the object of studying and researching the history of engineering and technology. It was named the Newcomen Society after the great inventor and its headquarters are at the Science Museum in South Kensington, London.

Thomas Newcomen

Thomas Newcomen was born in Dartmouth, Devon, in 1663/4. The exact day is not known but he was baptised on 24th February at St Saviour's Church. Little is known of his early education and training, but his father Elias was a merchant trader and a member of a small group of men responsible for bringing the well-known non-conformist scholar, John Flavell of Bromsgrove, Worcestershire, to Dartmouth in 1656. It seems more than likely that the young Thomas Newcomen received his early education from Flavell. This link with Bromsgrove is a very important one, because there was a Baptist group in Bromsgrove at that time. The Newcomen family had developed strong links with the Baptist faith, but there was no Baptist Chapel in Dartmouth at this time.

Elias Newcomen, Thomas' father, was a merchant trader in Dartmouth until his death in 1707. It is not known if young Thomas Newcomen served an established apprenticeship but it is generally thought that he established himself in business in his own right as an ironmonger in Dartmouth about 1685. The term 'ironmonger' at this time was very different from what we would think of today because they not only sold metal goods, but many of the items for sale were made on the premises. Newcomen's business activities

Fig. 11a: The references which form part of the Thomas Barney engraving which is preserved in the Birmingham Museum of Science and Industry. Note the handwritten note at the bottom.

REFERENCES

By Figures, to the several Members.

1. The Fire Mouth under the Boyler with a Lid or Door.
2. The Boyler 5 Feet, 6 Inches Diameter, 6 Feet 1 Inch high, the Cylindrical part 4 Feet 4 Inches, Content near 13 Hogsheads.
3. The Neck or Throat betwixt the Boyler and the Great Cylinder.
4. A Brass Cylinder 7 Feet 10 Inches high, 21 Inches Diameter, to Rarifie and Condense the Steam.
5. The Pipe which contains the Buoy, 4 Inches Diameter.
6. The Master Pipe that Supplies all the Offices, 4 Inches Diameter.
7. The Injecting Pipe fill'd by the Master Pipe 6, and stopp'd by a Valve.
8. The Sinking Pipe, 4 Inches Diameter, that carries off the hot Water or Steam.
9. A Replenishing Pipe to the Boyler as it wastes with a Cock.
10. A Large Pipe with a Valve to carry the Steam out of Iles.
11. The Regulator moved by the 2 Y y and they by the Beam, 12.
12. The Sliding Beam mov'd by the little Arch of the great Beam.
13. Scoggen and his Mate who work Double to the B. y, Y is the Axis of him.
14. The great Y that moves the little y and Regulator, 15 and 11 by the Beam 12.
15. The little y, guided by a Rod of Iron from the Regulator.
16. The Injecting Hammer or F that moves upon it's Axis in the Barge 17.
17. Which Barge has a leaking Pipe, besides the Valve nam'd In No 7.
18. The Leaking Pipe 1 Inch Diameter, the Water falls into the Well.
19. A Snifting Bason with a Cock, to fill or cover the Air Valve with Water.
20. The Waste Pipe that carries off the Water from the Pillon.
21. A Pipe which covers the Pistion with a Cock.
22. The Great Sommers that Support the House and Engine.
23. A Lead Cystern, 2 Feet Square, fill'd by the Master Pipe 6.
24. The Waste Pipe to that Cystern.
25. The Great Ballanc'd Beam that Works the whole Engine.
26. The Two Arches of the Great Ballanced Beam.
27. Two Wooden Frames to stop the Force of the Great Ballanced Beam.
28. The Little Arch of the Great Ballanc'd Beam that moves the No 12.
29. Two Chains fix'd to the Little Arch, one draws Down, the other up.
30. Stays to the great Arches of the Ballanc'd Beam.
31. Strong Bars of Iron which go through the Arches and secure the Chains.
32. Large Pins of Iron going through the Arch to stop the Force of the Beam.
33. Very strong Chains fixed to Pistion and the Plugg and both Arches.
34. Great Springs to stop the Force of the Great Ballanc'd Beam.
35. The Stair-Case from Bottom to the Top.
36. The Ash-hole under the Fire, even with the Surface of the Well.
37. The Door-Case to the Well that receives the Water from the Level.
38. A Stair-Case from the Fire to the Engine and to the Great Door-Case.
39. The Gable-End the Great Ballanc'd Beam goes through.
40. The Colepit-mouth 12 Feet or more above the Level.
41. The dividing of the Pump work into halves in the Pit.
42. The Mouth of the Pumps to the Level of the Well.
43. The Pump-work within the Pit.
44. A Large Cystern of Wood 25 Yards or half way down the Pit.
45. The Pump within the House that Furnishes all the Offices with Water.
46. The Floor over the Well.
47. The Great Door-Case 6 Feet square, to bring in the Boyler.
48. Stays to the Great Frame over the Pit.
49. The Wind to put them down gently or safely.
50. A Turn-Barrel over the Pit, which the Line goes round, not to slip.
51. The Gage-Pipe to know the Depth of the Water within the Boyler.
52. Two Cocks within the Pit to keep the Pump work moist.
53. A little Bench with a Bass to rest when they are weary.
54. A Man going to Replenish the Fire.
55. The Peck-Ax and Proaker.
56. The Centre or Axis of the Great Ballanc'd Beam. *that Vibrales 12 times in a Minute & each stroke lifts up 10 Gall: of water 51 yards perpend.*

must have been quite extensive because he bought large quantities of iron goods from the famous Foley ironmasters on the River Stour in Staffordshire and Worcestershire. In the year 1698-1699, a total of almost twenty-five tons was purchased from this one group.

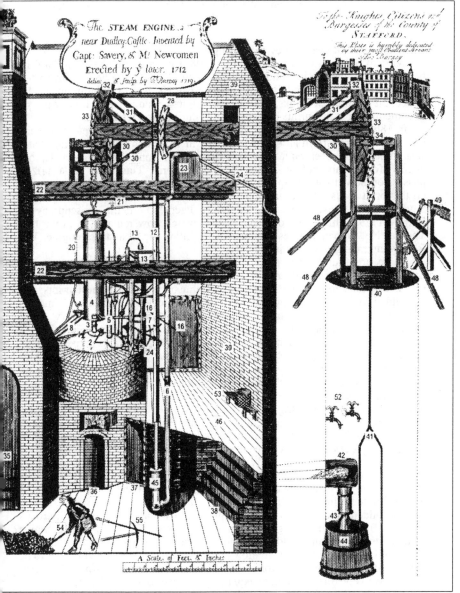

Fig. 11b: Thomas Barney's engraving of the world's first steam engine, drawn in 1719 with the reference numbers enlarged and clarified by the author.

Thomas Newcomen, who ran his ironmongers business at Dartmouth, had a partner, John Calley, whose trade was stated to be a plumber and glazier. In 1705, at the age of forty-two, Thomas Newcomen married

19

Hannah. the daughter of Peter Waymouth, a farmer from Malborough, near Kingsbridge, Devon.

The Discovery of the Barney engraving

William Salt was born in London in 1808. He was a member of a Stafford banking family whose local home was at Weeping Cross and he was interested in any material that shed light on Staffordshire's past. He built up a vast collection of printed books, pamphlets, original manuscripts, transcripts of documents and maps, prints and engravings during his lifetime.

In 1868, when this vast collection was being catalogued, a great discovery was made. A print from an engraving made in 1719 was found, proving that a steam engine had been built in 1712 to draw water from a coal mine at Tipton, twenty four years before James Watt was born! This print is a prized possession now displayed in the William Salt Library at Stafford. The caption on this print reads, 'The Steam Engine near Dudley Castle. Invented by Capt. Savery & Mr. Newcomen. Erected by ye later 1712 delin. & scup. by T: Barney 1719.' This is the only illustration which now remains to prove that Thomas Newcomen was responsible for inventing the first steam engine. The Barney engraving is shown in Fig. 11.

There are four known prints surviving from Thomas Barney's original copper engraving and all are preserved in the United Kingdom. Some very small variations are found in the references attached to each of these prints.

Added to the print discovered some time before 1876 by Samuel Timmins and now preserved in the Birmingham Museum of Science and Industry (print A), is a hand-written note stating that the engine, 'Vibrates twelve times a minute and each stroke lifts up ten gallons of water fifty one yards perpendicular.'

A second print (print B) is on display in the Science Museum, South Kensington, London. This print was bought from Messrs Quaritch by Dr Dickinson and presented to the museum in the name of the Newcomen Society. Both prints A and B have references to the engine printed and attached to the left-hand side of the drawing.

The other two prints are quite different. Print C, also in the Science Museum, London, and print D, in the Willaim Salt Library at Stafford, has the written references placed on each side of the engine details.

With some very careful study, it is apparent that all four surviving prints have been made from Thomas Barney's original 1719 engraving as small flaws appear across the fire door on all four prints. It is only the written references attached to the prints that vary.

The added reference to print A appears to be an afterthought. Perhaps Thomas Barney, on selling this print, was asked 'What does this engine actually do?'. He simply added the handwritten note which has been subject to very careful scrutiny because it allows the engine's original power output to be calculated accurately. The work done lifting 10 gallons of water 51 yards vertically, twelve times a minute, equates to a power output of 5.5 hp.

The print D in the William Salt Library has the words, 'Birmingham – Printed and sold by H. Butler in New Street'. An interesting contemporary reference to the Barney print has been found in Sweden where, in the journal of Swedish industrialist Jonas Alstromer (1685-1761), he states that in 1720 Thomas Barney sold these prints for two shillings.

Thomas Barney

Thomas Barney was a Wolverhampton file maker and artist. It appears that Barney had an acknowledged position in industry and commerce because Alstromer had met Barney at Wolverhampton to discuss the local mining and iron industry. However, Thomas's spare time must have been spent recording industrial events in the area and, when witnessing the erection of the world's first steam engine, he made a great effort to produce a copper printing plate so that he could produce prints of the engine for sale, giving Thomas additional income.

From the first recorded mechanical movement created by the expansive force of steam – Hero's Aeolopyle made in 100AD – a little over 1,600 years had passed and Thomas Newcomen was about to start the world's first steam engine to pump water from a coal mine at Tipton in South Staffordshire. In 1996, this is a very simple task using the pumps and motors that are commonly available, but in 1712 the lifting of water heralded the beginning of the industrial revolution.

CHAPTER 3

THE POTTERY PROCESS

Before describing the Dudley Castle engine in detail, I will describe another quiet revolution which has taken place. In 1980, a new method of pottery manufacture was being introduced into the ceramic industry and it can, without doubt, be classed as the industrial revolution of the pottery industry. It is doubtful if the miniature engines, around which all this research has been based, could have been made without this new manufacturing method.

Before I retired as Chief Development Engineer for Royal Doulton, I had been responsible for developing this revolutionary technique which incorporates the compacting of dry powdered clay at high pressure in precision steel dies. Since 1980, all the ceramic parts of the miniature engine houses, which includes bricks and tiles, have been made using this process. It is also very satisfying to realise that much has been discovered about the industrial revolution from the re-creating of these engines in miniature form.

GRANULATE PRESSING – THE QUIET REVOLUTION

Pottery, in its widest sense, includes all objects fashioned in clay and then hardened by fire. The basic method of forming clay into useful artefacts has remained unchanged for almost four thousand years. Clay is refined, first by removing the rocky fragments and then by both beating and kneading, and then adding water, to produce a material with a smooth texture that can be shaped and formed into useful objects. The clay, in this raw state, is generally referred to as 'plastic clay'. The water added gives the clay the ability to be re-formed.

The ancient Greeks and the Romans made many domestic items by forming the clay on to moulds made from terra cotta or earthenware, while plaster moulds had been used for a number of years in France, but the first major

Fig. 12: The first form of mechanised working of clay – a man-driven potter's wheel of the eighteenth century.

Fig. 13: By c1850 this type of machine was in everyday use in the pottery industry. With only minor refinements, these machines continued to be used until the early 1950s.

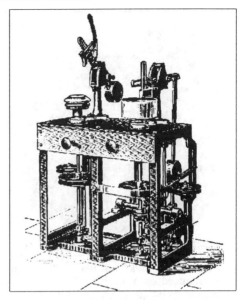

change came about in this country when, in 1743, Ralph Daniel of Corbridge introduced the use of the plaster mould into the British pottery industry.

The use of plaster revolutionized the forming process needed to shape the clay into intricate forms. The natural absorbency of this new material easily removed the water from the plastic clay, which greatly helped to release the formed object from the mould.

From the introduction of plaster into the pottery industry, there was an evolutionary development of the machinery and the materials which were used. Some of the earlier simple machines are shown in the Figs. 12 and 13. By 1965, the basic method of clay forming is shown in Fig. 14 and these machines, although still using wet plastic clay, became very complex. The machine shown here formed the domestic item by using a high speed rolling tool.

By the early 1970s, very sophisticated machines had been developed. The manufacture of flatware, such as dinner plates, saucers and side plates, was mostly automatic. One thing which had never changed was the clay which was used; this was prepared in large extrusion machines, or 'pugs'. It was still the basic material which was used by the early Greeks and Romans; in other words, plastic clay was still being used.

In the early 1970s, experimenters were busily working behind closed doors perfecting a technique which was to change all the ideas and the theories of clay forming. The large automatic machines soon became outdated and the new method of clay forming would change the whole pottery industry for many years to come. This new method used dry powdered clay, pressed at very high pressure, between steel and plastic dies and is now referred to in the pottery industry its 'granulate pressing'. I can hear the reader saying, 'What has this to do with the development of steam power in the eighteenth century?', but without this new forming technique it is doubtful if the project I set myself in 1974 could have come to fruition!

23

Fig. 14: A roller forming machine introduced c1950, now largely replaced by hydraulic presses forming the ceramic ware from dry powered clay.

Most of the bricks and tiles used to construct the miniature engine houses were produced on a machine designed and made by myself after 1980 using these new techniques. All the principles used in making the large hydraulic machines of the pottery industry were employed in miniature to make a machine capable of producing ceramic objects on a small scale.

The force needed to form a dinner plate is 500 tons, and the force needed to form the miniature bricks is 5.5 tons. It is very satisfying to think that making these miniature engine houses out of original materials produced using the very latest principles has allowed the writer to rediscover how the first steam engines were originally made at the start of the industrial revolution!

From the discovery that the earth had a pressure of 1 bar (14.7 pounds per square inch), it is perhaps appropriate to describe how this process works. The hydraulic pressure of these machines is 440 bar, almost 6,500 pounds per square inch. One of these enormous machines is shown in Fig. 15 pressing dinner plates.

The development of granulate pressing

The first experiment into the new forming techniques was carried out in 1966. At this date the factor limiting their use was the ceramic industry's inability to process the clay into a form which could easily be worked. These early experiments were made by crushing and grinding the clay into a powder. The granulometry of the powder made in this way was difficult to form between ridged steel and plastic dies; it simply would not flow into the intricate shapes of the dies.

The breakthrough needed took place in the early 1970s when the process of spray drying was introduced into the pottery industry. Spray drying was a well known method and had been used for many years in the food industry, but its introduction and use into the pottery industry really opened the way to bring this developing technique to perfection.

The clay is first prepared by blending the component parts which make up a clay or body recipe into a liquid form, generally referred to as a pottery slip.

Fig. 15: A 500 ton hydraulic plate forming machine, the principles of which form the basis for the design of the miniature brick making machine.

This slip is then pumped at a very high pressure into a heated tower 20 metres tall. The liquid clay enters this tower at the base. The slip pressure is just high enough to force the liquid to the top of the tower. The great heat inside the tower removes most of the water by evaporation and the clay particles then fall to the base of the tower by gravity. In this way, a spherical particle of clay is

25

Fig. 16: The specially prepared clay shown under a magnification of 38 times.

produced which flows very easily – as freely as sand in an egg timer. This free flowing of the particles of clay was to prove essential to this process. The clay prepared in this way is shown under high magnification above.

With the clay prepared, there was intense competition to develop the techniques and machinery which would be needed to form the clay into familiar domestic items.

The first machine to use the clay in this form was at the British Ceramic Research Association, Stoke-on-Trent, in 1971. This was only an experimental machine and not one that could be used to produce large quantities of domestic tableware. A substantial effort was needed, both in design expertise and financial backing, to quickly develop machines which were capable of producing pottery on a large scale. Sadly, as the financial backing was not forthcoming, this new machinery was not developed or made in England. The race to make the machinery which would be needed to process the clay in this way was taken up by one Italian and three German machine tool manufacturing companies. Now there are only two companies making these machines which are used all over the world and, I am pleased to be able to say, that I was involved in the development of these machines, spending a great deal of time in Germany perfecting the principle of operation for their use within the Royal Doulton group of companies.

It was on one of these visits to Germany in 1980 that I began to wonder if a small, simplified version of these machines would press the bricks which would be needed to construct my model engine houses.

How the process works

The clay is first of all prepared by spray drying into the spherical granular form shown in Fig. 16. The granules of clay are then pressed between two dies, one die, which is plastic coated, forms the underside of the plates; the upper side from which we take the food is formed by a profiled steel tool.

To press a clay powder uniformly, special conditions have to be met. The

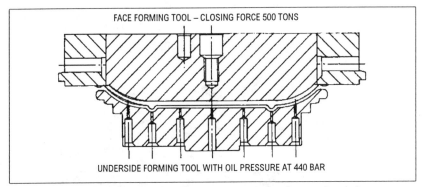

FACE FORMING TOOL – CLOSING FORCE 500 TONS

UNDERSIDE FORMING TOOL WITH OIL PRESSURE AT 440 BAR

Fig. 17: A cross-section of the plate forming process using dry powdered clay.

clay particles have to be transformed from their inert state into a workable solid form by the application of a uniform high pressure, and this is achieved by increasing the pressure of the oil behind the plastic die to slightly over 440 times the pressure of the earth's atmosphere, which is equal to 6.500 lbs per square inch. In technical terms, these machines are called isostatic presses.

With clay pressed in this way, the process of firing at a high temperature is always successful. The illustration Fig. 15 shows what these machines look like, while the detailed cross-section above shows how this principle works and shows a typical cross-section of this forming process. The item being made is 225mm in diameter and after the firing, it will have reduced to a little over 200mm in diameter.

The new forming process has been subject to controversy because in its infancy it was always referred to as 'dust pressing', but to whisper the word 'dust' in the pottery industry is fraught with danger and is synonymous with ill-health and the industrial disease of silicosis. A new name simply had to be found and it is now always referred to as 'granulate pressing'. Nowadays, a British pottery is always very clean, the word 'dust' is seldom heard, and silicosis is a disease of the past.

I do hope this diversion from the basic theme has been helpful, because it explains the evolution of the forming techniques which has been used to produce the miniature bricks.

The miniature bricks

The production of the miniature bricks has closely followed the evolution of the new forming technique. The first bricks to be needed were for the Boulton and Watt Lap Engine and about ten thousand bricks were produced. The traditional method of clay forming was used for this sectional building, and although very tedious, did provide me with good quality bricks.

The method used to form these bricks was the result of some very careful study. Many ways were experimented with, but the method which proved the most successful is shown in Fig. 18. In the time honoured way, the wet plastic clay was loaded into a small home-made extrusion pug. The pug was machined from a 150mm diameter hydraulic cylinder and the clay was then loaded into this cylinder and forced through a nozzle by means of a hand operated screw.

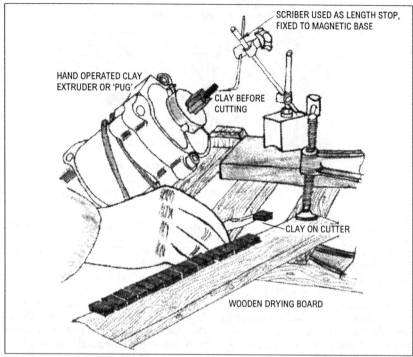

Fig. 18: Miniature brick making 'by hand'.

The size of nozzle on this device was the exact cross-section of the bricks. The extruded clay was then cut to the required length with a very carefully ground hacksaw blade. The length of each brick was measured and cut off when the clay reached the pointer shown in the illustration.

Wet clay is a very sticky substance and will adhere to almost anything. This property was put to good use because, after each small brick was cut, it fortunately adhered to the hacksaw blade and it was then easily removed on to the drying board. Consequently, during this transfer process there was very little distortion of the bricks.

The clay at this stage was still very wet, so the moisture content had to be reduced to about 5 percent if the subsequent firing was to be successful. For this to happen, the raw bricks were simply left in the sun to dry. They were then placed in a hessian sack and given a good shaking. After this treatment, the bricks really did look as if they had been made two hundred years ago!

The Firing

The small extruder, or pug, worked very well in forming the red plastic clay into the size and shape required to make the miniature bricks. They were made to the scale of one sixteenth the size of a common building brick which would have been in use in 1788. But another problem had to be solved if their visual appearance was to be the same as the bricks made over two hundred years ago. How were they going to be fired? The method in use during the

Fig. 19: Hand made bricks drying in the sun.

eighteenth century was to generate a great heat by the burning of coal in beehive kilns. It was this very process which was responsible for producing the vast range of colours seen in these older bricks.

This process of combustion requires a great amount of oxygen and a great heat. If the carbonaceous material in the clay is fused into each brick, the resulting appearance of the bricks ranges from bright red to almost black.

I am often asked how the bricks were made with so many different colours. It must be stressed that they are all made from identical clay, the only variation being in the amount of oxygen present during the firing or fusing process. It is the firing process which has produced this great range of colours.

There are only two atmospheric conditions in general use to fire clays into the solid material which we know as ceramics. This is the one described above and classified as reduction firing (which means reducing the oxygen present in the kiln).

The other process is to fuse the clay into the solid ceramic without generating the heat by combustion within the actual firing zone of the kilns. The heat to raise the kilns to the thousand or more degrees Celsius is provided by electrical heating elements. With the heat generated in this way, the oxygen is not removed from the firing zone and the objects are classified as being fired in an 'oxidizing atmosphere'. The electric firing burns away the carbonaceous matter from the clay to produce a brick of constant colour and size. This is why we now see so many new houses built from bricks all of the same colour.

It would be nice to be able to reproduce these early firing conditions and build our houses with bricks of many different colours and sizes, but clean air legislation prevents bricks from being fired by burning coal. However, after some careful thought, it was decided to fire the clay in a simple coal burning

Fig. 20: Firing the bricks for the miniature Lap Engine. Note the temperature measurement indicator.

kiln and to dismiss the use of the more sophisticated electric and gas fired kilns.

An oil drum was used for the coal firing, as can be seen in the photograph, Fig. 20 above. The lower 14" was cut away using an oxy-acetylene cutting torch. Several holes were cut at random around the sides and in the base and the drum was then placed on several concrete slabs in an uncultivated part of the garden situated well away from the house.

A refractory saggar was then placed on two bricks inside the drum. In the bottom of the saggar was placed a bed of coal, and on the top was placed the small clay bricks. The bed of coal was used to reduce the oxygen in the saggar, with the hope of producing the desired colour of bricks.

The saggar was 14" long by 9" wide and 6" deep. At each firing it could hold 4,500 bricks. Before the saggar was closed, a pyrometer was placed through a hole in the side which reached into the centre of the clay bricks. This was then connected to a temperature recorder which stood well away from the fire. The pyrometer was used to check when the correct temperature of 1050°C was reached. This method of temperature recording was not essential, but it ensured that two week's work was not ruined in a single firing! The saggar was now sealed by placing a refractory lid or 'batt' on top. A half a hundred weight of coal was now packed tightly around the saggar until the oil drum was quite full. Then the fire was lit with the aid of a propane gas burner, as can be seen in the photograph.

It took three hours to reach 1050°C before being held at that temperature for about one hour. The fire was then allowed to burn out and cool naturally.

Fig. 21: A saggar with 4,200 bricks after a temperature of 1050° Celsius has been reached.

The saggar was then removed and the miniature clay bricks were ready to use, made of the correct corresponding materials and fired just as they would have been in 1788. It must be remembered that during the drying and firing process, a shrinkage in size of some fifteen per cent takes place. This change in size has to be allowed for when making articles out of clay. The bricks can be seen in the saggar in Fig. 21.

This method was applied successfully to the small bricks used in the Boulton and Watt Lap Engine and to complete this engine, 10,000 bricks were made. What a daunting task it would have been without 'the quiet revolution' previously described. The remaining 130,000 bricks were produced using the new technique of granular pressing.

The Bricks Produced by the Granulate Pressing Method

When James Watt's Lap Engine model was completed, it was planned to research and construct Thomas Newcomen's Dudley Castle engine. The initial calculations indicated that over 42,000 miniature clay bricks would be needed to complete the model sectional building which would house the world's first commercial steam engine. In 1982, I was discussing the final design of the granulate presses which were to be introduced into the pottery industry and on one of the trips to Germany, my thoughts were centred upon the granular pressing – could my miniature bricks be formed by this new method? If this process could be mechanised, the production of the miniature bricks could be speeded up. It was now feasible to spend some time thinking about a design of a miniature version of the machines which are now in everyday use in the pottery industry.

The machine shown in Fig. 22 overleaf was the result of over 2,500 hours of trial and experimentation in bringing together the principles of producing powdered clay into a compact form ready to be heated to a temperature of over 1,000°C, so that the clay particles fuse or fire to form a solid ceramic body.

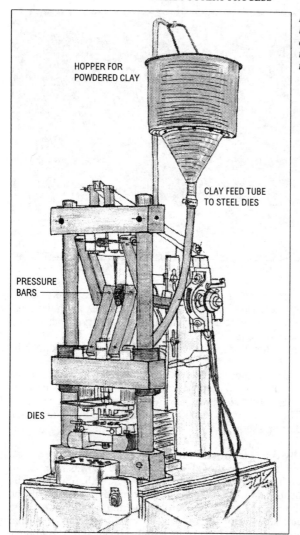

HOPPER FOR
POWDERED CLAY

CLAY FEED TUBE
TO STEEL DIES

PRESSURE
BARS

DIES

*Fig. 22: The miniature
brick making machine
designed and made to use
the latest clay forming
techniques.*

The clay granules are poured into the feed hopper. By the force of gravity, the clay flows into an automatic filler mounted over the high quality steel dies. With the die cavity full of powdered clay, the upper die or punch descends and compacts the clay with a force of 3 tons per square inch. With the application of this high force, the loose clay reduces to a thickness of about one half the original. With this reduction in volume, the clay is very strong and almost ready for firing, but the process produces very sharp edges on the bricks when they are removed from the ground steel dies. A quick solution to this problem was achieved by placing the pressed bricks into a hessian sack and giving them a good shake.

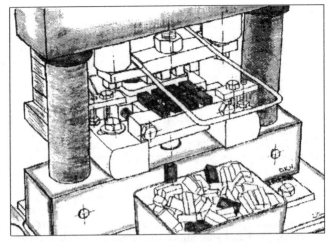

Fig. 23: An enlarged view of the miniature brick making machine showing the arrangement of the dies.

The arrangement of the dies positioned in the press are shown in Fig. 23. In this illustration, the press produces four complete clay pieces before being re-filled to repeat the process. When everything was working according to plan, it was very easy to complete six strokes a minute and, at this rate of production, 42,000 bricks did not appear to be an unrealistic target, but with the original method it was doubtful if the project of producing the miniature bricks would have been possible.

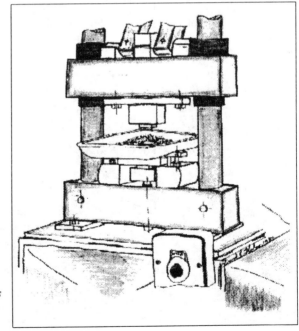

Fig. 24: The brick press set to make roof tiles from clay granulate.

The Small Brick Press

The 500 ton presses used to manufacture domestic tableware apply the force needed to compact the clay by the means of high pressure oil in large cylinders. The small press used to make the bricks closes with a very high force, the mechanical advantage being achieved by closing the press by the means of the toggle levers, seen clearly in Fig. 22.

The machine is operated by a one kilowatt motor driven through a high reduction gearbox, which in turn is connected to the toggle levers by a simple connecting rod. This operation very closely follows the principle of a piston and connecting rod within the internal combustion engine.

So that the clay can be compacted for a successful firing, a force of 5.5 tons has to be applied by this simple mechanical movement. The two vertical bars of this press are 40mm in diameter and, on the application of this force, actually stretch 1.5mm in length. The cross members of this machine do not appear to deflect or bend, perhaps this is understandable because they are made from 80mm square section mild steel. The brick machine was so successful at forming items from dry powdered clay that in 1995 a patent was taken out protecting its method of operation. Machines working on the principles developed from this press are now in everyday production making domestic tableware at the Royal Doulton works.

The press shown in Fig. 24 is all set to make the roof tiles which would be used to complete the miniature, but this time only pressing one item on each stroke. It did not seem appropriate to automatically fill the dies with powder, as only 3,000 tiles would be needed to complete the roof. The die was filled with clay granules by hand for each tile.

A New Method of Firing the Bricks

The bricks were easily made by the new machine. But what was now needed was a better way of firing them. The method used today is an accurately

Fig. 25: 42,000 miniature bricks compared with a handmade brick from the foundations of a house built in 1558.

controlled electric kiln, but the difference is that during the actual firing it burns coal.

Why should an electric kiln burn coal? The answer is to reduce the oxygen within the kiln to reproduce the conditions of a true coal fired kiln and to produce bricks whose appearance is similar to the bricks made almost three hundred years ago.

The coal is placed into the base of the saggar and its combustion burns away the oxygen and reproduces the vast range of colours found in the bricks made in the early years of the eighteenth century. The reason for these variations has already been described.

With the making of the bricks now completed, a start could be made on building the engine house. In Fig. 25, I show what a pile of 42,000 miniature bricks looks like. The hand-made house brick also shown in this photograph is from a house which was built in 1558.

I do hope this diversion into the development of the new manufacturing techniques within the pottery industry has proved interesting because, without this method, it is doubtful if an accurate scale model building to house Thomas Newcomen's engine could have been made.

Fig. 26: The prototype version of the model of Newcomen's Dudley Castle engine house, built to test new methods of construction.

CHAPTER 4

THOMAS BARNEY'S ENGRAVING AND THE MODEL ENGINE

The construction of the model of Thomas Newcomen's engine has been based upon very close study of Thomas Barney's engraving drawn in 1719 and required close consultation with the Newcomen Society to verify the interpretation of facts so that an accurate miniature engine could be built. As all the findings are based upon the references attached to this engraving (Fig. 11b), I thought it-would be appropriate to describe the evolution of the engine by using Thomas Barney's references (Fig. 11a) and describe what was going through scientists' minds at this time.

From the Heap of Bricks to the Engine House

The completed miniature engine at the chosen scale of $^1/_{16}$ full size shown in Fig. 28 actually required 42,200 miniature bricks and, with all the construction, research and the making of the many special tools, occupied all of my spare time over six years. For added interest, an approximate log was made of the actual time spent which, on completion, was found to total 6,246 hours.

Many weeks were spent calculating the building sizes and planning how to build the engine house. A careful study of the references on this print reveals some very important sizes; Reference 2 states that the boiler was 5 ft 6" in diameter and had a height of 6 ft 1"; the engraving states that the boiler was brought into the building through the double doors at the rear.

After many attempts to establish a practical layout for the building from the Barney engraving, a final design and size was chosen. This decision was difficult because, although the Barney print is very detailed, it is in no way drawn to scale. This can quickly be seen, for example, by the size of the fireman. He could not possibly reach and shovel coal on to the fire; and where was the coal pit mouth on this drawing? It is shown half-way up the gable end wall – the list of anomalies is endless. The intention was to make a model of Thomas Newcomen's 1712 Dudley Castle Engine as near as possible to how Thomas Barney had seen and drawn it in 1719 and it appeared that the only way that this aim could be achieved was to build a prototype to discover how the original building was constructed, make all the adjustments necessary and when completely satisfied, start all over again and build the final version.

This all sounds very obvious, but what it meant was that I had to build the model of the world's first steam engine twice! The first version of the building can be seen in Fig. 26 and it answered many questions. Was my chosen method of construction strong enough, bearing in mind that I was building a structure which would have huge sections of the walls missing so that people would be able to see inside? Would it simply fall down, and could it ever be moved without damage? As far as can be established, a miniature building with so many bricks had not been built before, but all my fears proved groundless and the engine house was then built again. This can be seen in Figs. 27 and 28.

Fig. 27: A view of the inside of the Dudley Castle engine model built by David Hulse.

Fig. 28: The outside of the completed model of the Dudley Castle engine.

The bricks used for the prototype model engine house were of uniform colour. The intention was to build this as quickly as possible, so a standard electric kiln was used without any atmospheric control and the bricks were fired in an oxidising atmosphere. However, the bricks for the real model were fired by the method described in chapter 3 with the atmosphere starved of oxygen, thus producing the colours of bricks which would have been seen in 1712.

Before construction finally began on the miniature engine, the Newcomen Society was consulted again to check whether any further information was available on this engine. Mr John Allen, an acknowledged authority on Thomas Newcomen, kindly provided me with the building drawings of the engine which was to be constructed at the Black Country Museum, Dudley and it was very satisfying to find that the drawing of the engine house only differed in the smallest details from my prototype model version.

A new perspective on the Barney Engraving of 1719

The Barney print in Fig. 11b is shown with his reference numbers enhanced by computer and I have included an enlargement of some of the details in Fig. 29. I will discuss each reference, though not necessarily in the order laid out by Barney. Each of his references will be illustrated with a pictorial view of the item and I discuss how each component was made in miniature. All the drawings which are to be shown relating to the engine have been taken from the model engine shown in Figs. 27 and 28.

Fig. 29: A much enlarged view of the boiler, cylinder and valve gear drawn by Thomas Barney, with his reference numbers computer enhanced.

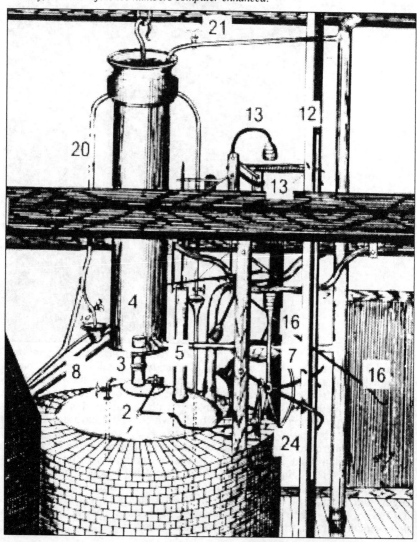

CHAPTER 5

THE BOILER AND CYLINDER DETAILS

Reference Number 1 (Fig. 30)
The Fire Mouth under the Boyler with a Lid or Door
The coal used to fuel the fire was shovelled through this opening, converting water into steam and setting into motion the world's first steam engine. Fire doors for boilers must have been made in their millions, but the one attached to Thomas Newcomen's 1712 Dudley Castle Engine is the first recorded one. The fire door on this engine would have been made of cast iron, set into the brickwork surrounding the boiler. On the miniature engine, the surround and the door are machined from solid pieces of mild steel. After being machined, a light grit blasting gave these components the appearance of having been cast in sand.

The fire door on Thomas Barney's engraving can be seen out of reach of the fireman. Barney has lowered the floor level to reveal more of the engine's details. The construction of the prototype model building helped to determine the working height of this wooden firing platform.

Reference Number 2 (Fig. 31)
The Boyler 5 Feet, 6 Inches Diameter, 6 Feet 1 Inch high, the Cylindrical part 4 Feet 4 Inches, Content near 13 Hogsheads
The steam which was condensed to allow the earth's atmospheric pressure to power the engine was raised in a vessel which derived its shape from many of the experiments which had been conducted by the earlier scientists, but the material this boiler was made from followed the design of the vessels which were in current use for the distillation of spirits.

The boiler's dome was made from a sheet of beaten lead and the base which was exposed to the burning coals was made from beaten copper. These two materials clearly indicated the technological limitations of the day; but

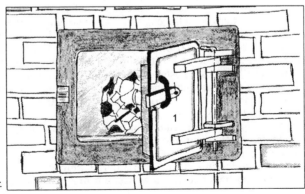

Fig. 30: The fire door.

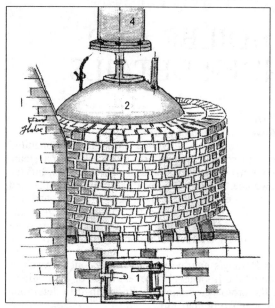

Fig. 31: The boiler.

choosing lead and copper does indicate Thomas Newcomen's connection with the mining industry of Cornwall, as both copper and lead were to be found there in abundance.

The choice of these two materials put a limitation on the steam pressure which could be safely used to achieve a water and steam tight joint between the lead dome and the copper base, while the use of these two materials is a good indication of the steam pressure which was most probably used. It was unlikely that the boiler could have worked at a pressure exceeding two pounds per square inch. This large vessel was not used to raise steam to a high pressure but to generate a large volume of steam which could be condensed to create a vacuum against which the earth's atmospheric pressure could act.

The boiler was completely surrounded by brickwork – the heat from the burning coal was only allowed to act upon the base of the boiler and not on the outer edge – as was used on subsequent boilers. Applying the heat in this way put strict limitations on the engine, as not enough steam was generated to keep the piston in a continuous cyclic mode – this will be described in full later.

It was found on later engine installations that much more steam could be generated by exposing a greater surface area to the burning coals. On the 1712 engine, the heat was wasted by being drawn into the chimney stack before transferring the maximum amount of heat into the water contained within the boiler positioned above. A large amount of potentially useful heat was thus lost to the outside atmosphere.

Perhaps the very delicate joint between the lead dome and the copper base had forced Newcomen into using this design. He had to keep the heat away from this joint because, with the melting point of lead only being 327 degrees Celsius, joint failures would have occurred quickly and frequently.

With lead being such a low strength material, the boiler of the miniature

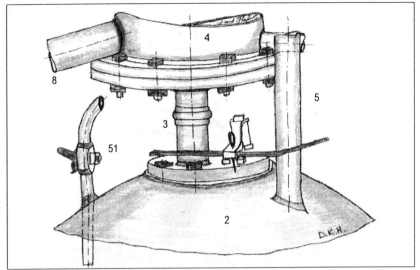

Fig. 32: The top of the boiler with its connection to the cylinder.

was made from mild steel sheet, beaten and riveted together. This was then covered with lead sheet to give the required appearance. The complete miniature boiler was then set into a circular wooden former. This former was then completely surrounded by miniature ceramic bricks to give the appearance which is indicated on Thomas Barney's print of 1719.

The mortar bonding the miniature bricks together, a ceramic tile adhesive, was applied with a medical hypodermic syringe.

Reference Numbers 3 & 51 (Fig. 32)
The Neck or Throat betwixt the Boyler and the Great Cylinder
The Gage-Pipe to know the Depth of the Water within the Boyler

Helped by natural convection, the steam rose at a low boiler pressure into a 3" diameter vertical standpipe and then into the large brass powering cylinder positioned above. This brass stand pipe was an integral part of the flange which was bolted into position and formed the top of the boiler.

Operated by the lever shown in Fig. 32, a wiping valve or plate moved in a circular arc, cutting off and then turning on the flow of steam into the cylinder above. This valve was held into position by the steam pressure in the boiler.

The origin of this valve closely follows the design of the steam valves used on Thomas Savery's 'Miners Friend'. Perhaps Thomas Newcomen simply copied a tried and tested method on this engine.

On the Dudley Castle engine, the cylinder base, together with the standpipe and boiler flange, were most probably cast in brass and as a single piece. On the miniature engine, three brass pieces were silver soldered together to form this component.

Reference 51 is a very important addition by Thomas Barney, because this standpipe would only work with the boiler under slight pressure. The lead

43

Fig. 33: The cylinder.

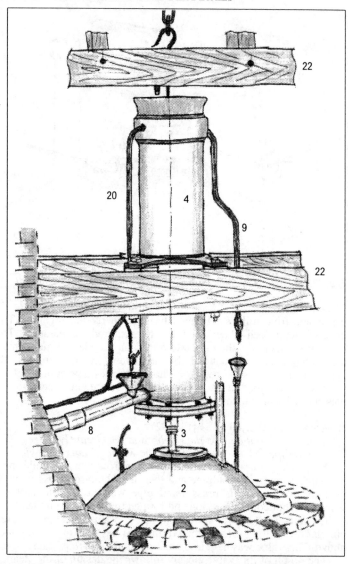

dome of this boiler was not just there to direct the steam by natural convection into the cylinder above, but also used a slight steam pressure.

The standpipe was used to gauge the water level in the boiler and, when turned into the 'on' position, either water or steam would issue forth. If steam issued, this indicated that the water level was below the stand pipe; and if water issued, this indicated that all was well. The water level of this boiler was critical because, if this level fell below the joint between the lead dome and the copper base, the heat in the copper base would quickly rise and possibly melt the soldered joint between the two halves. The normal working temperature of the steam and water was 100 degrees Celsius, and if this temperature was allowed to increase, a major rebuild would have to take place. The standpipe would have been made from brass and secured into position by a wiped plumber's joint.

Reference Number 4 (Fig. 33)

A Brass Cylinder 7 Feet 10 Inches high, 21 Inches Diameter, to Rarifie and Condense the Steam

Let us pause and reflect before reading further technical details of this engine, because we are looking at a vertical brass cylinder containing a piston which is compelled to move against a vacuum which was created by condensing the steam from the boiler below. The earth's atmospheric pressure of 14.7 lb per square inch moved this piston through the 21" brass cylinder's bore. All subsequent steam engines, even the internal combustion engines powering today's cars, contain pistons moving on the inside of cylinders.

The value of Thomas Newcomen's invention is that it presented mankind with a powering force which could be developed into the machinery which we all know and use today. It is a very sad that Thomas Newcomen did not receive the credit in his lifetime for so great an invention.

This really was the first piston, moving on the inside of a cylinder and it was linked by chains to the overhead rocking beam to create a mechanical movement. With the piston in the uppermost position, the steam was allowed to fill this cylinder and, when completely full at the low boiler pressure, a valve was turned to shut off the supply of steam from the standpipe 3.

At the base of this cylinder was positioned an additional pipe through which cold water was allowed to spray. This cold water spray, on contact with the heated steam, caused condensation and the formation of a vacuum; it was this vacuum created on the underside of the piston which allowed the earth's atmospheric pressure to force the piston to make one movement and thus complete a working stroke of the engine.

It is generally accepted that the 7 feet 10 inches size of this reference relates to the actual stroke but this is not clear and could refer to what is stated as the overall height of this cylinder.

News of this great invention soon spread in this country and abroad, and many scholars and diplomats came to see the engine at Tipton pumping water from the mine workings one hundred and fifty feet below the surface. But they were only allowed to see the outside of the engine house. Thomas Newcomen was closely guarding the engine inside the building; he wanted to protect its method of operation and gain the full commercial benefit for himself and his family.

Marten Triewald was a noteworthy engineer who, in 1716, came from Sweden and studied the principles of how the Newcomen engine worked. He is the only person known to have worked with both John Calley and Thomas Newcomen. In 1734 he wrote an account of the development of the steam engine in England between the years 1716 and 1726, in which he states that on one occasion the Spanish Ambassador at the English court came from London with a number of foreign visitors to see the engine at Tipton, but Thomas Newcomen would not allow them to see the inside; the only thing which they were permitted to see for themselves was the pumping effect from the outside of the building. They were definitely not allowed to see what was creating this great mechanical movement. They are said to have retired to London in a bad temper with their mission unfulfilled.

Captain Savery's name included on the title of the Barney print does at first seem puzzling when he played no part in this engine's development. The answer is that Thomas Newcomen never took out a patent on his engine. A patent, or privilege as it was then called, had been granted to Captain Savery for the construction of the fire engine, later it would become known as his 'Miner's Friend'.

In 1698 Savery was granted a patent No. 356 for 'A new Invention for Raising of Water and occasioning Motions to all Sorts of Mill Work by the Impellent Force of Fire, which will be of great use and advantage for draining mines, serving towns with water and for the working of all sorts of mills where they have not the benefit of water nor constant winds.' This patent covered England, Wales and Berwick-on-Tweed for the usual 14 years; in 1699 it was extended for 21 years, running until 1733. Newcomen used this patent.

Many of the references on the Barney print relate to the piston on the inside of the brass cylinder, but none state or show how this piston was made. The most probable design could have been copied from the glands or seals of a large hand pump. These used a 'bend of leather', clamped between stout metal plates, holding the leather seal into a cup shape. This type of piston would have taken into account that the cylinder of this engine, was hand-finished. The only finishing which was possible after casting in 1712 was the removal of all the flashing and sand texture, using hand files and scrapers. With a diameter of twenty-one inches, there would have been just enough room, to hand file and scrape this cylinder.

The Large Brass Cylinder

The casting of this twenty-one inch diameter and eight feet long brass cylinder was an unique event in 1712. What is surprising is that there is no record stating for certain either the foundry's name or what products they actually made. The foundry must have had the skill and expertise to be able to cast large items because, assuming this cylinder had a wall thickness of one and a half inches, the Newcomen cylinder weighed about 1.25 tons.

Knowing the size and also the weight of this item might suggest two possible foundries. The first foundry capable of melting this large volume of metal would be a cannon foundry, but most cannons cast in the early eighteenth century were made of iron. The second foundry capable of handling this size cylinder would seem the most probable as they had become used to casting large items in non-ferrous metals. They were, of course, bell foundries. Brass parts were supplied for Newcomen engines at about this time by a brass founder from Bromsgrove, William Brookes. What is not known is whether this foundry was capable of casting such a large item as the cylinder, but the foundry's location at Bromsgrove would suggest that the cylinder may have been cast there because, in 1712, an item of such weight would have been difficult to transport over any great distance. At this juncture, I contacted a foundry in North Staffordshire who specializes in the casting of non-ferrous metals, J T Price of Chesterton. Arthur Stevenson showed me how they cast cylinders today (1994) by a method which he considers would not have changed over the past three hundred years. Perhaps the Newcomen cylinder

Fig. 34: Derek Johnson of Price's Foundry at the core making machine.

was cast using the following method. The most important feature of Thomas Newcomen's cylinder was to have an internal surface which would seal and work without precision machining. In 1712, precision boring machines were not available. Arthur also showed me how the company makes the core which in turn reproduces the texture, and final accuracy, of their cylindrical castings.

The core is built around a central spindle by spiral layers of natural hemp. With the hemp securely tied into position, a layer of muck sand is applied over the hemp base. The whole assemblage is then placed upon a support and turned by hand. It is made to spin as accurately as possible with the aid of wooden formers and metal cutters. After this treatment, the spindle, hemp and muck sand are placed into an oven to dry.

With the drying complete, an additional layer of hemp is again wound around the core and then another layer of muck sand. The whole process is continually repeated until the desired diameter of the core is achieved. From a core made in this way, Arthur assured me that the surface finish of the casting's

Fig. 35: A completed core.

bore could not be bettered. As I stood and watched the moulder at work, this came as no surprise when the composition of this primitive mix was revealed. The four ingredients were black casting sand, well rotted loam and an unspecified helping of horse manure, all mixed into a workable paste with water. All these ingredients were readily available and could have been used in 1712. Why horse manure? This material is very fibrous and, when the molten brass comes into contact with a core made in this way, the fibres are burnt away. This allows the molten metal to solidify without minute blow holes which could spoil the finished surface. The gas produced by cooling is allowed to escape into the core and out to the atmosphere. Modern moulding materials do have one added advantage – the mould maker does not have to change all his clothing each evening before home-time!

Fig. 36: The buoy which controlled the engine.

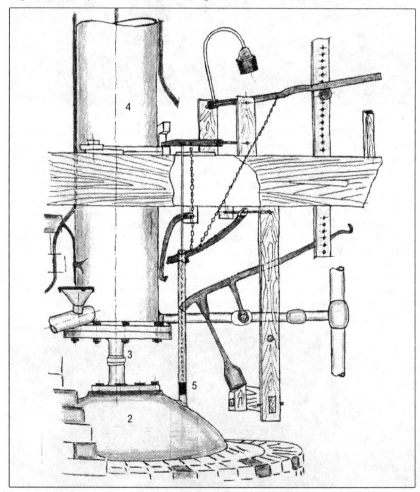

CHAPTER 6

ENGINE OPERATING MECHANISMS

Reference Number 5 (Fig. 36)
The Pipe which contains the Buoy, 4 Inches Diameter
This reference shows how many of the component parts are fitted together to perform the automatic opening and closing of the control valves. The valves enabled this engine to be maintained in a 'near cyclic operation'. Many myths surround the valve gear which operated the earliest Newcomen engines. One very popular but unsubstantiated one is that a young boy called Humphrey Potter was controlling an engine without an automatic valve operating mechanism. He is said to have devised a system of rods and levers, all moved by the massive overhead rocking beam, to open and close the valves to convert a manually operated engine into one working automatically. Although this is a very interesting account of these engines, it cannot be true because, as can be seen from Reference 5, the engine is shown in an advanced state and was already fitted with an automatic operation when Thomas Barney illustrated this engine in the early eighteenth century.

It must have been Thomas Newcomen who perfected this type of operation in the many years he spent conducting experiments, possibly in Cornwall. When he assembled this engine in the Midlands, he already knew how to make the engine work in a continuous cyclic way, or was the engine made to run automatically between the years 1712 and 1719, *before* Thomas Barney produced his famous engraving?

The earlier statement of the engine running in a 'near cyclic operation' is thought-provoking, because the early boiler powering Thomas Newcomen's engine simply could not raise enough steam to keep the engine running continuously at the stated speed of twelve pumping strokes per minute. The engine would run at this stated frequency until the steam was exhausted. The engine would then stop until the boiler had raised enough steam for the engine to continue until again running out of steam.

The controlling mechanism for this operation was the buoy sliding up and down on the inside of the 4" diameter vertical standpipe, shown as reference 5. The engraving does not state what material the 4" diameter pipe was made from, but with John Calley being an experienced plumber, the most probable material would have been lead sealed to the top of the boiler with a 'wiped joint'.

The buoy would fall in this pipe when the steam was not at a high enough pressure to run the engine and, in this position, the control latches would be held in the off position. When the water boiled and the steam pressure rose, the buoy would rise and disengage the control latches and, through gravity, the weight attached to the other end of the lever would turn the stop valve into the on position and the engine would be set into motion again. When all the available steam had been used, the pressure would decrease allowing the buoy to fall in the standpipe and this would hold the catches in the off position,

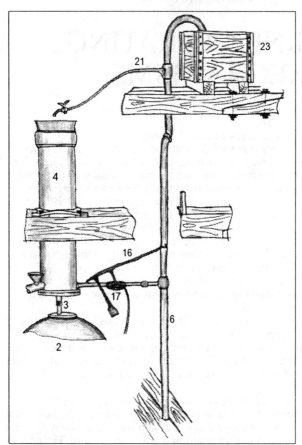

Fig. 37: The master pipe which delivered water from the well to the header tank.

stopping the engine. The engine would then remain stationary until enough steam had been raised to again lift the buoy.

Reference Number 6 (Fig. 37)
The Master Pipe that Supplies all the Offices, 4 Inches Diameter

Beneath the engine house and directly below the master pipe 6 was constructed a brick lined well, called the 'in house well'. Before discharging the water lifted from the mine workings, the water flowed into this well and only flowed to waste when enough water had accumulated to satisfy all the requirements of the engine. The master pipe 6 was the delivery pipe from a pump positioned at the bottom of the well.

This automatic valve operation and automatic water supply substantiates the claim that Thomas Newcomen must have perfected these requirements on a full size engine before venturing to the Midlands. He could not have known what was actually required by experimentation with models and it is unfortunate that his earliest experiments are not documented.

The master pipe lifted water to the header tank supported on the upper great sommer beams. The most likely material for this 4" pipe was lead, with the individual supply pipes attached by wiped joints.

Reference Number 7 (Fig. 38)
The Injecting Pipe filled by the Master Pipe 6, and stopped by a Valve

Water flowed along the horizontal pipe 7 and, on entering the 21" diameter brass cylinder, formed a cold spray which was directed into the cylinder full of steam. Possibly one of the greatest discoveries of all time, it was found that on condensing the steam a vacuum was created on the underside of the piston.

The vacuum then allowed the earth's atmospheric pressure to push down and move the piston to the other end of the cylinder. This completed a powered stroke for this engine. The heavy pump rods which were attached to the other end of the massive overhead rocking beam then pulled the piston seven feet ten inches back to the top of the cylinder, which would then allow another vacuum cycle to start and again draw the piston to the other end of the cylinder. This was repeated several times until the boiler ran out of steam.

The formation of a vacuum by this water spray was what the scientists and engineers had been working almost 1700 years to achieve – perhaps it is little wonder that its discovery by a relatively unknown man from Dartmouth sent waves of utter disbelief through the academic world.

There are two widely held theories about how Thomas Newcomen made this discovery, but the following theory is seen as most probable. It is thought

Fig. 38: The water injection pipe 7 and the barge 17.

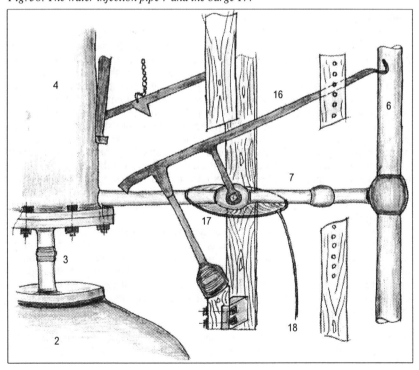

that Thomas Newcomen was experimenting with a piston and cylinder, surrounded by a jacket of cold water. An imperfection in the inner cylinder wall had previously been repaired by soft solder. The heat from the incoming steam melted this solder and allowed water to spray into the steam contained within the inner cylinder. A vacuum was quickly formed which drew the piston down the bore of the cylinder.

Reading through Marten Triewald's 1734 account of his association with Thomas Newcomen and John Calley leaves the reader in no doubt that Triewald was not a man of great modesty. On no account was he going to give the credit to the inventors for this discovery and actually said: "Though somebody might think that this was an accident, I for my part find it impossible to believe otherwise than that what happened was caused by a special act of providence. To this conclusion I who knew personally the first inventors have been brought more than ever considering that the Almighty then presented mankind with one of the most wonderful inventions which has ever been brought into the light of day, and this by means of ignorant folk who had never acquired a certificate at any University or Academy."

It would have been nice if the discoverer of this phenomenon had received the credit for it in his lifetime.

The 4" diameter horizontal pipe 7 was thought to have been made of lead and enters the base of the large brass cylinder, turning through ninety degrees so that the water spray could project vertically as far as possible into the steam filled cylinder and create the desired effect of condensation.

Reference Number 16 (Fig. 38)
The Injecting Hammer or F that moves upon its Axis in the Barge 17

The discovery of the useful force created by the condensation of steam now had to be regulated. A means had to be found to switch on and off this new found phenomenon on each cycle of the engine, and the engine completed twelve cycles or working movements in each minute. The injection hammer shown as 16 was used to control the water spraying into the cylinder. The movement of this lever was controlled by the combined efforts of the Scoggen (see reference 13) and the up and down movement of the sliding beam 12. The water spray was switched off by the sliding beam, pushing the injection hammer into a position so that is could be restrained by the catches shown in the drawing. The water spray was started by these catches being released, allowing the hammer to fall by gravity until its fall was stopped by coming into contact with a leather covered anvil. This combined action on the lever or hammer turned the control valve through ninety degrees for each working stroke of the engine.

This injection hammer was forged from wrought iron. Collieries employed blacksmiths who could have made this unusual item, either close to or on the assembly site. The valve moved by this lever was most probably of a conical design and made of brass. This type of valve could be easily made and had been used for many years in the brewing industry. But never before had this design been used to control the motion of a steam engine, where three separate requirements had to be met.

Firstly, to control the flow of cold water, followed by containing the steam which was to be condensed to form the vacuum. The injection hammer on the miniature engine was cut from a solid piece of mild steel and is lightly forged

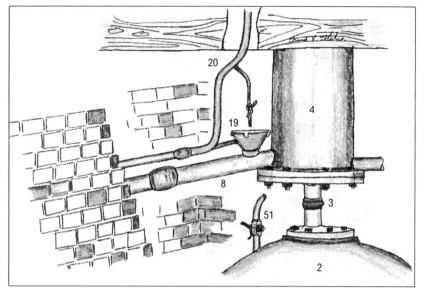

Fig. 39: The 'sinking pipe' or drain which carried off the hot water and steam from the cylinder.

so as to have the appearance of having been made almost three hundred years ago. The control valve is made of brass and is very similar to a design used on the engine in 1712.

Reference Number 8 (Fig. 39)
The Sinking Pipe, 4 Inches Diameter, that carries off the hot Water or Steam

The creation of a vacuum by the means of a water spray must have given Thomas Newcomen and John Calley a great deal of satisfaction; there had been so many years of trials and experiments. Marten Triewald says in his 1734 account that Mr Newcomen worked at this firemachine for ten consecutive years on what was needed to harness the potential power and make the engine work repetitiously. The Sinking Pipe 8 allowed the accumulation of water from the spray and also the build-up of water from condensing the steam to drain away from the large brass cylinder. The pipe was fitted with a valve 19 which allowed the accumulation of water to drain into the in-house well. This drainage was successfully carried out when the piston returned to the top of the cylinder, before starting another powering stroke of the engine.

All these contrivances were practical solutions to problems which had not been encountered before and, for each solution, credit must be given to the problem-solver from Dartmouth.

Reference Number 9 (Fig. 40)
A Replenishing Pipe to the Boyler as it wastes with a Cock

Marten Triewald, in his 1734 account of meeting and working with Thomas Newcomen, says it had taken ten years to perfect the formation of a vacuum by spraying cold water into a steam-filled vessel. What is not mentioned is how

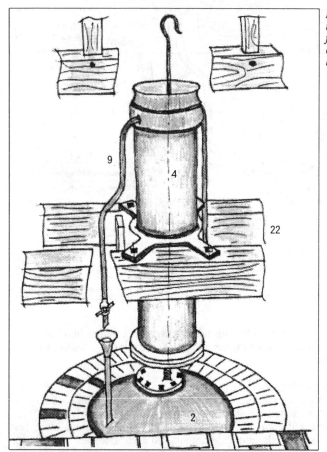

*Fig. 40: The
replenishing pipe
from the top of the
cylinder to the
boiler.*

long Thomas Newcomen spent on his experimentation, that is, on converting this idea into the full sized engine shown in the Barney engraving. The time which was spent can only be speculation, but the working engine pumping water from the coal mine at Tipton must have taken a great effort. For example, the simple looking pipe leading from the top of the cylinder is an automatic water feed to refill the boiler with water lost through the production of steam. The top of the piston was flooded with water to produce an airtight seal and this water was fed by a small pipe from the master pipe 6.

The good piston seal was achieved and maintained by allowing much more water than was needed to flow above the piston. The surplus water then flowed through the down pipe 9 to re-feed the boiler below. On entering this standpipe, fitted with a funnel, this head of water overcame the low internal steam pressure of the boiler which was refilled by gravity alone.

Both the standpipe and the pipe fitted with the cock, or control valve, were made of lead and held into their respective positions by a traditional plumbing joint – possibly all done by John Calley.

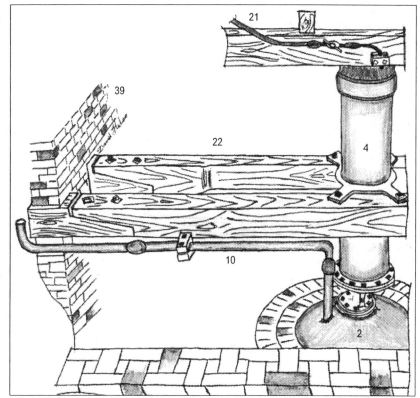

Fig. 41: The pipe 10 was used as a primitive safety valve.

Reference Number 10 (Fig. 41)

A Large Pipe with a Valve to carry the Steam out of Door

The large pipe 10 must have been fitted for a dual purpose. The main reason was to maintain the steam at a constant pressure – this would enable the powering cylinder to be repeatedly filled with steam and ensure that a good vacuum was formed after condensation. The adjustment of the valve ensured a constant back pressure which allowed the engine to work in a predictable way. The pipe and valve also acted as a safety device, which could prevent the steam pressure rising and damaging the delicate joint between the lead dome and the copper base of the boiler.

Before this pipe and safety valve were fitted by Thomas Newcomen, many calamities must have occurred. A rise in steam pressure would also increase the temperature of the metals of the boiler, also bringing the soldered joint even closer to a melt down. If this joint failed, the engine would be out of working order for a very long time because the soldered joint was completely surrounded by two courses of brickwork which would have to be removed and rebuilt again after the repair had been completed.

Thomas Barney does not give a reference on his engraving as to how the boiler was initially filled with water. I have given this problem considerable thought and have concluded that the safety pipe, 10, could have been used to deliver the 650 gallons of water into the boiler. Could the pipe have had this third function?

Reference Number 11 (Fig. 42)
The Regulator moved by the 2 Y y and they by the Beam 12

The steam was produced by boiling 650 gallons of water. This amount of water half-filled the lead and copper vessel. Contemporary references say that about half the internal space of the boiler must be filled by steam. A simple calculation reveals that the internal volume of the hemispherical lead dome equalled a space that would have held 650 gallons. This space had to be filled by steam at a low pressure before the engine could be started to pump the water from the mine workings.

The control valve 11 was positioned at the top of the lead dome. This simple valve was operated by levers attached by linkages and all were repositioned by the up and down movement of the great balanced beam. To start the engine into motion, this valve had be set in the off position. This would allow the boiling water to raise steam to the low pressure which was needed to create the vacuum and run the engine. With the piston in the raised position,

Fig. 42: The controlling linkages for the regulator 11 are described as the "2 Y y", a description which becomes obvious from the shape of the components!

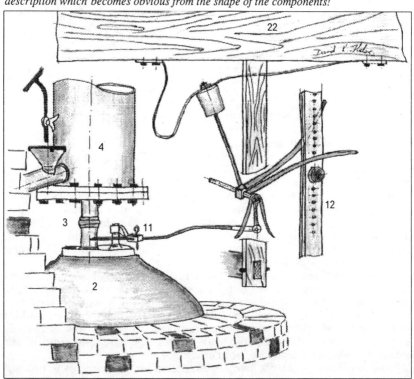

the valve 11 was moved into the on position by hand. This would allow the 21" diameter and almost 8 ft long cylinder to be filled by steam and, when completely full, the valve 11 was turned by hand into the off position. With the cylinder now sealed and full of steam, the water spray was introduced, the valve positioned midway along the horizontal pipe 7 being turned by hand.

The cold water spray, on contact with the hot steam, caused condensation and the formation of a vacuum. This vacuum then drew the piston through the bore of the cylinder. Now with the beam 'in the house', to use an eighteenth century expression, the pump rods then pulled the piston back to the top of the cylinder before another stroke could commence.

This action was repeated several times by hand before engaging the catches which would enable the valve to be operated automatically by the rocking movement of the great balanced beam.

The design of this control valve, set inside the dome of the boiler, appears to be a tried and tested method to control the flow of steam. It is first seen controlling the steam output on Thomas Savery's engine of 1698 and this must have been a very successful arrangement because the steam supply to all cylinders of atmospheric engines was controlled in this way for the following seventy years. During this time, almost eight hundred engines were made.

The valve was a simple brass quadrant, moving in a circular arc in order to cover or uncover the full bore of standpipe 3. The valve was successful and simply because of the two brass components rubbing together which wore their faces into a reliable seal. The two faces were held together by the internal pressure of the steam in the boiler and the valve could work for a long time with little maintenance.

This type of valve arrangement continued to be used until the late eighteenth century and was only replaced with the introduction of the pickle-pot condenser on some later engines.

Reference Number 12 (Fig. 43)

The Sliding Beam movd by the little Arch of the great Beam

The drawing (shown overleaf) shows an assemblage of parts that has to be ranked as the biggest single step in the history of engineering! Here is a vertical brass cylinder with a piston moved by the earth's atmospheric pressure, the steam being controlled by the up and down movement of the sliding plug rod 12 which moved the levers attached to the steam and water supply valves. Not only is this the world's first commercial steam engine, but it is also working *automatically*!

Never before in recorded history had an inventor taken such a unique thought and converted it in one single step into a working machine. In 1712, the only accepted form of mechanised working was by wind and water power and, on a much smaller scale, in horology. Thomas Newcomen's engine was not the product of steady evolution; he had not copied the idea from any of his predecessors. Newcomen presented mankind with a force which could be used to drain the mines of water and so allow the rich seams of coal and minerals to be removed without the constant threat of flooding. More importantly, he proved that a mechanical movement and a powerful force could be created by heat and steam. Subsequent scholars and engineers could now refine this idea and provide steam engines which would eventually be used in all industries.

The sliding beam 12 is attached by chains to a curved beam, or little arch head. This beam is secured into its working position on the overhead rocking beam by wrought iron straps and bolts.

The chains securing the sliding beam or plug rod into position are placed with one at the top and the other four feet below this beam. One chain pulls the beam up and the other chain pulls the beam down. This simple idea converts the radial movement of the overhead beam into true linear motion to accurately control the levers and the valves situated below the arch head.

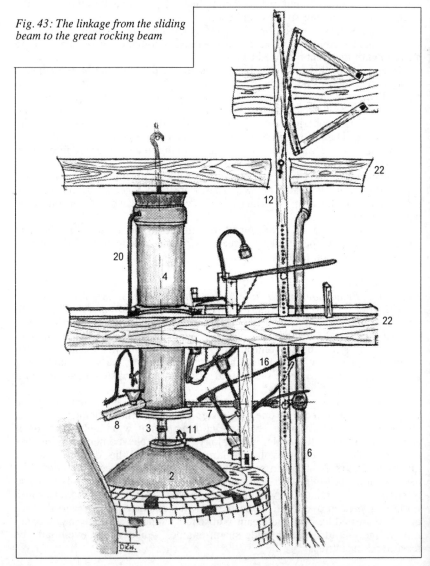

Fig. 43: The linkage from the sliding beam to the great rocking beam

At the lower end of this sliding beam is attached a long wrought iron rod. This rod extended about ten feet below the actual engine house floor and, at its lower end, operated the in-house well pump. This was the pump used to supply the engine with all its water for the production of steam.

The sliding beam was most probably made from English oak and, as can be seen in the drawing, the tappets operating the valve control levers are positioned on each side, located in holes drilled right- through the beam. The tappets were positioned by trial and error until the engine was running efficiently.

Reference Number 13 (Fig. 44)
Scoggen and his Mate who work Double to the Boy, γ is the Axis of him

The wording of this reference does at first seem odd, because what Thomas Barney is trying to say is that this mechanism is operated by two separate entities.

With the water level in the boiler correct and producing enough steam to run the engine, the 'buoy' or small piston located on the inside of the vertical stand pipe moves up and down with the changing steam pressure, releasing the catch shown in Reference 7. This allows the injection hammer to fall and rest upon the leather-faced stop shown in this drawing.

When the water is not boiling and there is not enough steam to run the engine, the catch holds the hammer in the raised (off) position. This stops the engine until enough steam has been raised for it to continue to pump water.

As shown in Reference 13, the use of the word 'double' means that this

Fig. 44: The "Scoggen" which controlled the opening of the valve to start or stop the engine. This drawing should be studied in association with Fig. 36 where the device can be seen from the opposite side.

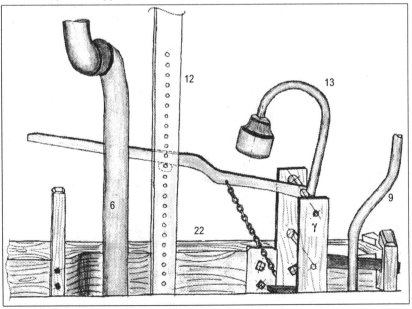

catching arm is also operated by the raising and lowering of the sliding beam 12. A tappet positioned at the side of the beam is holding the lever in the raised position. The use of the words 'Scoggen and his Mate' must hold a unique place in the history of steam power, because the word Scoggen does not appear to have any known use or meaning outside the description of this mechanism! It warrants an entry in the Complete Oxford English Dictionary which states:

SCOGGEN, of obscure origin, as it was a quasi-personal name. An automatic contrivance for opening valves in Newcomen's steam engine c1713).

The weight attached to the end of the odd-shaped arm ensured that the long lever was firmly held against the wooden stop. With the lever on this stop in a horizontal position, the engine could continue with a pumping stroke or movement. All contemporary references to this valve operating mechanism state that cords were used, not chains as I have shown. Chains were used because the miniature engine had to be assembled on the workbench before being permanently positioned inside the engine house. A cord would not have lasted very long and, when it eventually did break, a repair would have been almost impossible. Hopefully, the chains will continue to work without any repairs and so demonstrate how this valve gear operated in 1712.

Some very careful planning and thought has to go into miniature engines where some components are a structural part of the building. Every part has to be complete before careful assembly inside the confines of the model structure. Once in position, this part itself becomes a permanent fixture which could be difficult to adjust or modify.

Reference Numbers 14 and 15 (Fig. 45)
The great Y that moves the little y and Regulator, 15 and 11 by the Beam 12
The little y, guided by a Rod of Iron from the Regulator

The sliding beam, reference 12, was raised and lowered by the action of the great overhead rocking beam. Positioned at the lower end of this sliding beam

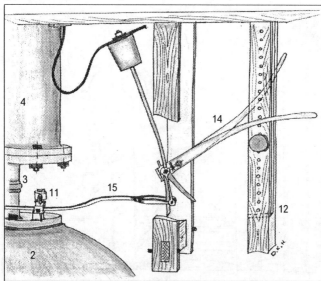

Fig. 45: Now the 'great Y' and the 'little y' become clear! This ingenious mechanism gave a snap action to the regulator valve controlling the steam entering the cylinder.

was a series of holes with two tappets, one each side of the beam. The levers were moved into their working position shown in the drawing by the main beam moving 'out of the house' – when the piston was at the top of the cylinder.

When the main beam had completed its movement, at the top of the piston stroke, the sliding beam reversed and tripped the great lever. The tumbling bob, shown with the leather strap attached, fell over the central position, giving lever 15 a sharp knock to move regulator 11 to the off position.

The action of these levers opening and closing regulator 11 is worthy of some thought, because it is only when this mechanism is fully understood that the reader can appreciate what a clever device this really is.

What is taking place is conversion of the slow movement of the sliding beam into a snap action which is required to control the flow of steam into the vertical brass cylinder. The tumbling bob moves slowly into the vertical position and falls by gravity to open quickly and then to close the steam supply regulator.

This method of controlling the supply of steam into the cylinder was so successful that the principle of operation remained in use on atmospheric engines unchanged for the next seventy years.

The complete assemblage of levers shown in this drawing all worked successfully, taking into account the variable movements this engine made pumping water from the mine.

The row of holes in the sliding beam allowed the tappets to be repositioned to suit the variable pumping requirements of the engine. When positioned close together, the tappets reduced the engine's pumping ability to a minimum, while when positioned at each end of the row of holes, the engine would pump to its maximum capacity.

The adjustment controlling the angular movement of the tumbling bob was also very clever. A leather strap stopped the bob's movement when the regulator 11 had been either opened or closed. This belt was extended or reduced in length to suit the variations which would have occurred when constructing the engine into a structural part of the building.

All these levers on the miniature engine were hand forged and the weight on the tumbling arm was cast from lead. The one deviation from the 1712 materials was the leather strap – this was replaced by modern polyurethane. Leather would have perished with time and would have been difficult to replace inside such a confined space.

Some idea of the miniature engine's size can be gauged by the pivot pin connecting lever 15 to the regulator 11. This pin is just 0.018" in diameter.

Reference Number 17 and 18 (Fig. 38)

Which Barge has a leaking Pipe, besides the Valve namd in No 7
The Leaking Pipe 1 Inch Diameter, the Water falls into the Well

These references on Thomas Barney's print of 1719 can only be appreciated with long and careful study of Fig. 38. Reference 17 is shown only as a faint outline of a drip tray which must have been designed to catch the water escaping from the main control valve positioned midway along the injection pipe 7. Great difficulty must have been met making this valve water and steam-tight and still able to move easily from the 'on' to the 'off' position. It must have been easier to catch the escaping water than to try to make a valve which was watertight! The technological limitations of 1712 must have been the controlling factor for the use of this type of control valve – a precision engineered sealing arrangement could not have been made for this early engine.

The drip tray, or 'Barge' as it is referred to, would have been made from a beaten sheet of lead. This material would have been easily made into the intricate shape needed around this valve. The pipe 18 can be seen draining the water into the inhouse well. This pipe would also have been made from lead, easily attached to the Barge by the means of a plumber's joint.

Reference Numbers 19 and 20 (Fig. 46)
A Snifting Bason with a Cock, to fill or cover the Air Valve with Water
The Waste Pipe that carries off the Water from the Piston

If an atmospheric engine is to work at its most efficient and before any condensation takes place, the great brass cylinder must be completely full of steam. A special valve was designed to achieve this and it was fitted to the base of the cylinder 4, shown as Reference 19.

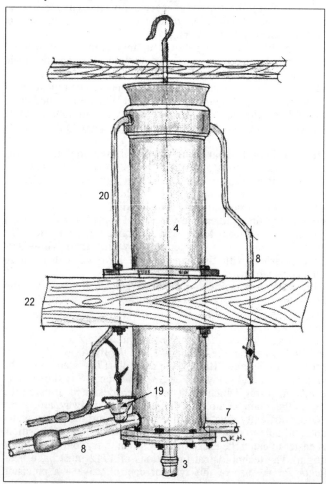

Fig. 46: The pressure release valve and its sealing arrangements.

With the piston rising before the powering stroke and drawing the steam into the space beneath, a slight variation in pressure would occur. This variation would prevent the cylinder from becoming filled uniformly with steam. The special valve 19 overcame the pressure variation and ensured all was well before condensation. Put simply, this was a pressure release valve which would allow an excess of pressure to escape and was set to allow the engine to run as its best; the valve also allowed all the air present within the cylinder to escape to the atmosphere, before being completely replaced by the incoming steam. Inside the funnel shown in the drawing, fig. 46, was a conical seated poppet valve, held on to its sealing face by gravity. To ensure a complete seal and retain the created vacuum, water was allowed to flood over this valve from the vertical pipe 20. From this vertical pipe can be seen a smaller pipe fitted with a stop valve, which could be used to regulate the flow of water into this funnel. When the valve was completely submerged, the excess of water flowed away to waste, or into the in-house well.

So successful was this valve in retaining the created vacuum and maintaining the steam inside the cylinder at a constant pressure that its design remained unchanged for almost a hundred years and it continued to be used on atmospheric engines until the early 1800s. It only ceased to be used when it was no longer needed because of the introduction of engines working on steam produced at a high pressure. It is very difficult to appreciate that the solutions to these engineering problems were unique and were not in any way based upon previous experience. Never before Thomas Newcomen, or since, have design ideas remained unchanged and worked successfully for such a long period of time!

The valve was most probably made from brass and sealed into a housing, also cast from brass. It was joined to the cylinder base and the drain pipe 8 by a wiped plumber's joint.

Reference Number 21 (Fig. 47)
A Pipe which covers the Piston with a Cock

The water which flowed through this pipe, reference 21, performed three very important functions, all needed to keep the engine in constant motion.

The primary function was to have a constant flood of water over the piston's top surface. This produced a good airtight seal between the leather piston seal and the cylinder wall. Without this constant flood of water, a seal between these sliding surfaces would have been impossible to achieve. In 1712, the inner surface to the cylinder had only been made smooth by hand files and scrapes and would not have been accurate enough to achieve a reliable seal by the more sophisticated sealing methods used on future engines, which had cylinders made accurate by machines designed and made for boring cannons.

More water flowed through pipe 21 than was needed for this method of sealing. The excess was used to refill the boiler with the water that was lost through evaporation or the production of steam. This flowed vertically down through the standpipe 9 to refill the boiler by gravity.

Reference Number 22 (Fig. 47)
The Great Sommers that Support the House and Engine

These great wooden beams were most probably made from the best English oak and were cut to an 18" square section. Being built into the two gable walls,

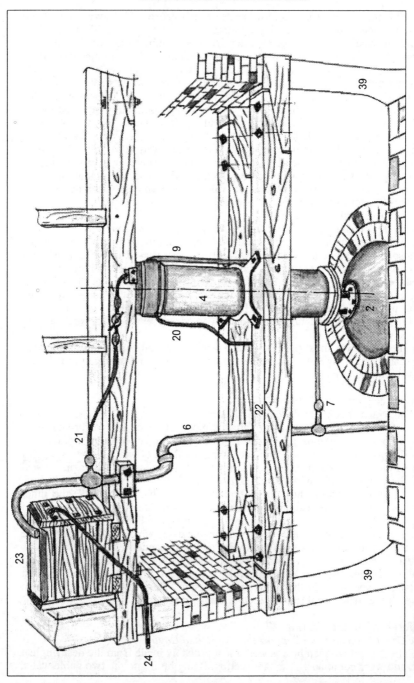

Fig. 47: This drawing shows the cistern 23, the sommer beams 22 and the pipe 21 which delivered the water which sealed the piston to the cylinder wall.

they became a very important structural part of the engine house. They gave the whole structure immense strength and also provided a strong platform on which to attach by bolts the principal part of the whole invention -the 21" diameter brass cylinder.

In 1712, the heavy baulks of timber would have been cut to their approximate size by two men using a pit-saw. The final size would be achieved by using an adze to remove the slivers of wood.

To enable the miniature engine to be tried and tested outside the confines of the engine house, it was decided to install these beams into the structure by the traditional joints shown in the drawing opposite. In 1712, each beam would have been a single piece spanning the full width of the building.

Reference Numbers 23 and 24 (Fig. 47)
A Lead Cystern, 2 Feet square, filld by the Mafter Pipe 6
The Waste Pipe to that Cystern

The header tank, with a capacity of about fifty gallons, was supported by the two upper sommer beams. The water entered the tank by the master pipe 6. The function of this tank is almost identical to tanks positioned in our own lofts which are used to keep the pipes full of water and free from air locks – water stored in the header tank was also used to ensure that the master pipe remained full at all times. The surplus water flowed to waste through the overflow pipe shown as Reference 24.

The reference on Thomas Barney's engraving states that this tank was made of lead, but the one made and fitted into the miniature engine house is made of wood and fitted with a lead liner. If the model tank had been made completely of lead, it would not be very strong and could easily be damaged, so the timber has been added to give it extra strength.

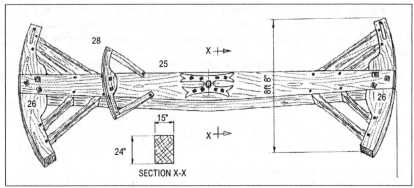

Fig. 48: The 'Great Ballanc'd Beam that works the whole Engine'.

Fig. 49: The preventer springs 34 and iron rod 32 seen at the top of the wooden frame outside the engine house.

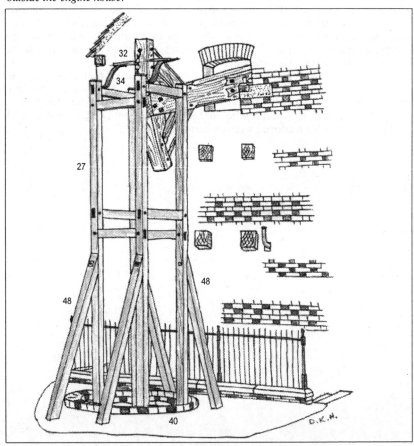

CHAPTER 7

THE ENGINE HOUSE AND PUMP DETAILS

Reference Numbers 25 & 26 (Fig. 48)
The Great Ballanc'd Beam that Works the whole Engine
The Two Arches of the Great Balanced Beam

Thomas Newcomen used the overhead rocking beam to pump water from the mine workings and this proved to be so suitable that this method of pumping water was in use for the whole period of mine drainage, and was used by both atmospheric and high pressure steam engines.

Pumping engines were in continuous use for the following two hundred and forty five years. The last beam engine to be built and used was in Redruth, Cornwall, where an enormous engine with a 90" diameter cylinder pumped 27,000 gallons of water each hour from the mineral mine workings 1,700 ft below the surface. This engine was made in 1891 and continued pumping water from the Taylors shaft until 11.30 am on the 28th September 1955.

The engine made by Thomas Newcomen had a main beam made from English Oak with a weight of 1½ tons while the engine at Taylors shaft was made by Harvey & Company of Hayle and had a beam cast in iron with a weight of 52 tons.

The beam of the 1712 engine was made from a single piece of oak, 24 ft in length with a maximum cross-section of 24" deep and approximately 15" wide. Thomas Barney's engraving of this engine shows that the beam had a parallel upper face and was tapered on the underside.

At each end of this great beam, bolts secured two very large, curved baulks of timber with a radius taken from the central pivot point. These became known as 'arch heads' and were used to compel the large operating chains to move in a straight line. This straight line movement was essential for smooth running when the chain was connected to the piston rod.

At the other end of the great beam, chains were connected to rods to operate the pumps positioned in the mine shaft.

The operating beam was very strong and was made using the traditional woodwork practice of mortise and tenon joints, securely locking the arch heads into position by the use of wooden dowels.

The size of this beam for the construction of the miniature engine was evolved by first building the engine house complete with the boiler and the large brass cylinder. The distance was then measured from the gable end wall, where the beam was going to pivot, to the centre of the main cylinder. This dimension was then doubled to establish the length of the beam.

Much care has to be taken when miniaturizing wooden items because, although it is very easy to reduce the scale of full size items, the grain cannot be scaled – wood with a much closer grain has to be used. The Japanese Oak tree is very slow growing and, when well seasoned, gives the right appearance of English Oak but closer to the correct scale of the miniature engine.

Fig. 50: This photograph shows the importance of using an alternative wood in scale models. The texture of Japanese oak is shown on the left, with English oak which would have been used in 1712 shown on the right.

Reference Numbers 27, 32 & 34 (Fig. 49)

Two Wooden Frames to stop the Force of the Great Balanced Beam
Large Pins of Iron going through the Arch to stop the Force of the Beam
Great Springs to stop the Force of the Great Ballanced Beam

Positioned at the upper edge of each arch head was a strong wrought iron bar which was used to prevent any damage by the piston being drawn too far towards the boiler or, at the other extreme, the piston actually being drawn out of the cylinder. One of these iron bars can be seen positioned through the arch head in the drawing where the beam is shown through the end gable wall. In this drawing, the bar is shown in contact with the strong springs which are secured by bolts to the wooden support structures.

On the inside of the engine house was a similar arrangement, but this time it was mounted on the two upper sommer beams. The function of this mechanism was to be a safety device which would prevent any major damage in the event of a chain failure, either on the piston end of the beam or within the mine shaft. If either chain broke, the beam would be halted by the contact with the springs and so prevent the piston being drawn into the boiler. If the failure was on the piston rod, the stops would prevent the beam from being drawn out of the house and into the mine shaft.

Devices very similar to Thomas Newcomen's design were fitted for almost two hundred years to all future beam engines. James Watt actually called them 'preventers' when they were fitted to his engines.

Reference Numbers 28, 29, 30 & 31 (Fig. 51)

The Little Arch of the Great Ballanc'd Beam that moves the No. 12
Two Chains fixd to the Little Arch, One draws down, the other up
Stays to the great Arches of the Ballanc'd Beam
Strong Barrs of Iron which go through the Arches and secure the Chains

To give a parallel action to the up and down movement of the sliding beam 12 required a new solution from the method first used to provide a parallel movement where the chain 33 was connected to the piston rod. If the sliding beam 12 was to have the force to move the valve operating gear, this beam had to be both pulled up and down. In addition to the force required to change the position of the control valves, an additional force would be needed

Fig. 51: Details of the Great Arches and the Little Arch at the end of the Great Balanced Beam.

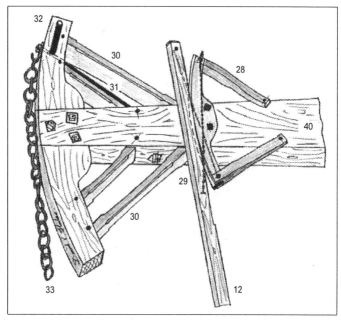

because this sliding beam extended well below the engine house floor level and also operated the pump positioned at the bottom of the in-house well.

This pump lifted water through the standpipe 6 to supply the engine with

Fig. 52: A photograph of the end of the great balanced beam made from close grained Japanese oak inside the model engine house.

all that was needed to run continuously. The Little Arch 28 was a very practical solution to this problem. Two chains of equal length were then connected to each end of this radiused beam. Their other ends were then firmly secured; one at the top and the other lower down to the sliding beam 12. For the miniature engine it was found necessary to have a tenon locating in a groove at the back of the sliding beam 12. Without locating these two components together in this way, a reliable movement could not be guaranteed.

With all these neat and practical solutions, it is perhaps not surprising that Thomas Newcomen was treated with incredulity by academia. The linear action by chains to a rising and falling rod also continued to be used until 1797 when the Derbyshire engineer Francis Thompson erected a rotary engine and connected the piston rod to the overhead rocking beam by the use of chains. This double action engine used two cylinders, one above the other, operating on the principle developed by Thomas Newcomen in 1712.

Reference Number 33 (Fig. 51)
Very strong Chains fixed to Piston and the Plugg and both Arches
Of all the component parts which were made to assemble the steam engine at Tipton in 1712, the great arch head chains would have caused the least problems, because where the engine was sited was the centre for all chain manufacture. This area of England made chains both large and small for use worldwide.

Although Thomas Barney says that these chains were very strong, the force exerted by the engine was relatively small and, if this engine was running well, the maximum force would not have exceeded 3,500 lbs. The chains, made from wrought iron with each link welded in the forge, would have been small when compared with the chains used for anchorage of shipping.

The chains for the miniature engine were also hand made, but this time each link was carefully silver soldered and chemically blackened to disguise the shiny appearance of the silver soldering.

Reference Numbers 35, 36 & 37 (Fig. 53)
The Stair-Case from Bottom to the Top
The Ash-hole under the Fire, even with the Surface of the Well
The Door-Case to the Well that receives the Water from the Level
From the main floor level, where the engineman would have been able to make all the working adjustments to the engine, a flight of eighteen stone steps descended into the basement. This flight of steps can be seen in the drawing beneath an arch which was actually built into the chimney stack. The basement was directly under the man shown about to stoke the boiler with coal.

The ashpit appears to have had enough room to enter the space beneath the fire bars. This is shown as reference 36. The basement must have been a very hot and dark place, because the only lighting was the glow from the burning coals and a little candle power. When you know that the door, reference 37, led into the in-house well, I could not be tempted through this door with only candlelight for illumination. The steps on the model descending into the basement were all cut from millstone grit. This stone was chosen because of its fine texture and each step was cut to the required size using a diamond slitting saw. The door leading into the in-house well was made from Japanese oak,

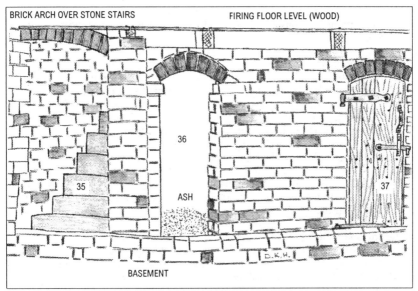

Fig. 53: The basement and its construction. The door 37 led to the in-house well.

Fig. 54: The wooden firing floor and staircase 38. What a fire risk this must have been!

with all the metal fittings made by hand forged mild steel.

To complete the basement area, the floor was covered with slabs of micaceous sandstone.

Reference Number 38 (Fig. 54)
A Stair-Case from the Fire to the Engine and to the Great Door-Case

Seven feet above the stone floor of the basement was positioned a wooden platform, from which the coal was shovelled into the furnace. A flight of six steps allowed the engineman to have easy access from the main floor level 46. What these steps looked like can only be speculation because they are not shown on Thomas Barney's engraving of 1719 – the steps made for the miniature engine are typical of the early 1700s and are shown as reference 38.

The wooden platform. referred to as the firing floor level, would cause today's factory inspector many problems. Just imagine – a wooden floor under a firing door with all that heat! To prevent the floor from burning, a continuous supply of water must have been at hand.

Reference Number 39 (Fig. 55)
The Gable-End the Great Ballanc'd Beam goes through

If ever a sight was synonymous with early mine pumping engines, a large overhead rocking beam protruding through a gable end wall must be it. This configuration, first used on the Newcomen engine in 1712, was to be a daily sight pumping water from mine workings for the following two hundred and forty-five years. There could be no more economical way of supporting the main pivot bearings than securing them centrally into the end gable wall. No additional support had to be designed or made and the main beam was rigidly

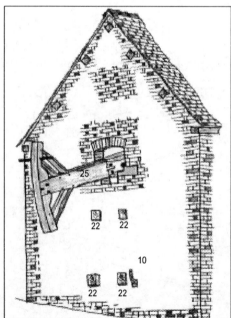

Fig. 55 left: The gable end wall through which projected the great balanced beam.

Fig. 56 below: Flemish bond brickwork.

Fig. 57: The coal pit mouth, showing the windlass and the wooden support frame.

held into its working position. The only added expense was that the gable wall, through which the beam protruded, was usually built thicker and stronger.

This gable wall can be seen with the bricks laid with one stretcher followed by a header that alternates on every row. This type of bonding is known as Flemish bond. How thick this gable wall was built in 1712 can only be speculation; the model engine is constructed with a wall of three bricks thickness.

Through the end gable wall can be seen the large support timbers for the roof of the engine house. These supports, constructed of 12" square sections of English oak, not only support the weight of the roofing tiles but also tie the two end walls together and give much added strength to the whole structure.

The details for the roof of the miniature engine house are taken from the measurements of a barn at Chirk Castle near Oswestry. These details were measured and sketched before being translated into the miniature scale of one sixteenth. The tiles on the roof are completely made to scale and actually do hold on to the laths by the traditional two nibs. The handmade ridge tiles are then firmly cemented into position.

Reference Numbers 40 & 41 (Figs. 57 and 58)
The Colepit-mouth 12 Feet or more above the Level
The dividing of the Pump work into halves in the Pit

The opening into the mine working through which the water was drawn was of almost eight feet in diameter. This was a brick-lined shaft extending to the full depth of approximately one hundred and fifty feet. The wording of reference 40 does need careful thought, because the twelve feet which is referred to is the vertical distance of the opening to the coal pit mouth, i.e. ground level, which is positioned above the in-house well.

73

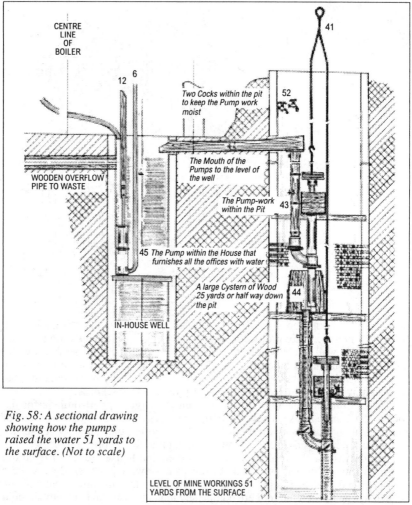

CENTRE
LINE
OF
BOILER

12 6

41

52

Two Cocks within the pit
to keep the Pump work
moist

WOODEN OVERFLOW
PIPE TO WASTE

The Mouth of the
Pumps to the level of
the well

The Pump-work
within the Pit   43

45 The Pump within the House that
furnishes all the offices with water

A large Cystern of Wood
25 yards or half way down
the pit   44

IN-HOUSE WELL

*Fig. 58: A sectional drawing
showing how the pumps
raised the water 51 yards to
the surface. (Not to scale)*

LEVEL OF MINE WORKINGS 51
YARDS FROM THE SURFACE

A wooden launder was connected to the outlet of the mine pumps before directing the flow of water into the in-house well. This wooden launder was also twelve feet below the coal pit mouth. The reason why these early pioneers divided the pumping rods into two is easily explained. In theory, it would have been very easy to position one pump at the bottom of this well and have a pipe ascending one hundred and fifty feet to the surface. But in 1712 this would have been impossible, because the only vertical pipes which could be used were made of wood. Elm was usually the chosen material but, when using wood, careful thought was necessary because the head of water at one hundred and fifty feet would have created a pressure which could have burst these vertical lift pipes. At one hundred and fifty feet, a pressure of 75 psi would have been exerted by the force of gravity alone.

By dividing the rods and operating two separate independent pumps, one positioned at the bottom of the coal pit and the other halfway down, the pressure on these wooden pipes had been reduced from 75 psi to a pressure of $37^{1}/_{2}$ psi, well within the limit of safe working for these wooden pipes.

The wooden pipes were made by the use of augers or on machines originally developed for boring cannons, with each length of hollowed out timber secured with a spigot. They were held together by wrought iron straps and bolts. The drawing showing the pumps within the eight feet diameter shaft illustrates how these pipes were connected before the water was forced from the mine workings. to be discharged via the in-house well to the out-flow from the mine.

Every function of Thomas Newcomen's engine and its application to mine drainage brought new problems and, for each problem, a neat and practical solution was found which could be made by local craftsmen using the technology of the early 1700s.

Reference Numbers 42, 43 & 44 (Fig. 58)
The Mouth of the Pumps to the Level of the Well
The Pump work within the Pit
A Large Cystern of Wood 25 Yards or half way down the Pit

The Thomas Barney drawing of 1719 and many other excellent drawings made in the early 1700s only record what these early engines looked like above the ground level. There is no detail to show how the pumps were arranged within the mine shaft and on no drawing does it show what design of pumps was used with these early engines.

The pumps within the mine shaft were drawn by the author from details gleaned from a manuscript written in 1743 by J T Desaguliers FRS. This is a well detailed but rare manuscript entitled *A Course of Experimental Philosophy*. Many types of pumps are shown, but plunger type pumps would have easily fitted into the eight feet diameter vertical pit shaft and were most probably used to raise the water from the coal mine at Tipton.

The plunger pump was invented in about 1660 by Sir Samuel Morland and worked by drawing the water into a primary chamber, which was filled on the pump's negative stroke. The positive stroke then closed a flap valve at the base of the pumping chamber and redirected the water vertically through another flap valve and into the large wooden cistern positioned midway down the mine shaft. As stated previously, this reduced the water pressure by a half on the wooden standpipes. From halfway down the mine shaft, an identical plunger pump to the one at the base of the pit was used to force the water vertically into the wooden launder, before flowing into the in-house well.

These pumps were suitable for forcing water vertically through the stand pipes because they were very easily adjusted to give whatever water pressure the different mine depths demanded.

When a mine needed to be drained but was of an unknown depth, the pumping pressure required could be ascertained quite easily by trial and error. These pumps later became known as 'deadweight pumps'.

The main body of the pumps would have been made of cast iron, with a plunger most probably of brass. The deadweights were at a later date cast in iron but in 1712, to reduce the deadweight's physical size, lead could have been used. However, the use of weights on these pumps can only be speculation because of the lack of below ground details. It is not possible to make and show these pumps in their correct position on the model engine. Above the ground level, everything is to a scale of one-sixteenth full size, but whatever can be done with a mine shaft which was originally one hundred and fifty feet deep? Only one pump is shown on the model, at the top of the coal pit.

The large tank of wood, reference 44, was to receive water from the lower pump and would have needed a capacity of about fifty gallons. This size, with a little left in reserve, would be large enough to draw out the water from the mine successfully.

Reference Number 45 (Fig. 58)
The Pump within the House that Furnishes all the Offices with Water

The in-house well is shown in Fig. 58 with a small pump which is operated by the great overhead beam, raising and lowering the wooden sliding beam 12.

This small pump was identical in operation to the large main well pumps, but the water which was forced vertically into the standpipe 6 was only needed to replenish the water lost through the production of steam in the boiler. All the water raised from the mine workings flowed into the in-house well. It was only when the pump, reference 45, had supplied the engine with water that the surplus flowed away to waste.

Positioned almost at the top of this well is shown a wooden pipe, through which the surplus water finally flowed after being pumped from the mine workings one hundred and fifty feet below.

Reference Numbers 48, 49 & 50 (Fig. 57)
Stays to the Great Frame over the Pit
The Wind to put them down gently or safely
A Turn-Barrel over the Pit, which the Line goes round, not to flip

When a 1990s civil engineer has to lift heavy weights such as those equal to the weight of the pumps within the Tipton mineshaft, it would be easy to telephone the local plant hire company. All the equipment would be delivered to the site and the job would be completed very quickly. However, when this engine was being assembled, things were very different. Everything needed for the assembly and maintenance of the pumps also had to be designed and put together on the site. The windlass shown in Fig. 57 is typical of the eighteenth century, with two operating handles turning the wooden roller, around which is shown a hemp rope. The windlass shown does not have a ratchet to hold the load in position as one of the operating handles would have been taken off the squared shaft and reversed, thus preventing the barrel from revolving freely by being pressed against the wooden structure of the windlass.

The rope from the windlass was wound around the large pulley, referred to by Thomas Barney as a 'turn barrel' then descended into the mine shaft. The whole wooden support structure over the shaft had to have some additional strength to withstand the extra load of lowering the pumps by the windlass. This extra strength was given by the four wooden supports, shown as reference 48.

Reference Number 52 (Fig. 58)
Two Cocks within the Pit to keep the Pump work moist

Positioned above the wooden drainage launder inside the coalpit mouth are two water taps. There is no obvious use for these taps - the only clue is given in the wording of reference 52 and they are illustrated in Fig. 58.

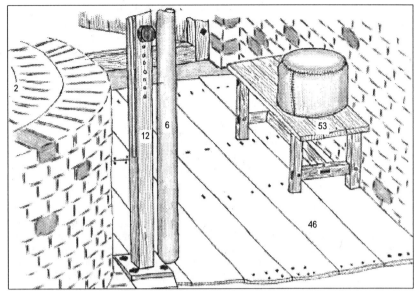

Fig. 59: The floor over the well and the engineer's seat.

With wood used for many of the drainage parts such as the wooden lift pipes and the wooden launder, these taps could have been used to ensure these wooden items did not shrink – perhaps there were long periods when the engine was not in use. A hose could have been connected to direct water to the most vulnerable parts. This purpose does seem probable because domestic wooden barrels used to catch rain water must always be wet to prevent shrinking and leaking. The drawing does not show where the water supply to these taps came from.

Reference Number 46 (Fig. 59)
The Floor over the Well

This was the main floor level for the engine house and is where the attendant would have stood to make all the adjustments which would have been necessary to maintain the engine in continuous motion. How many personnel the engine required to operate it is not known. The wooden floor shown above is a suspended floor which covers the in-house well described earlier.

The sliding beam 12 and the master pipe 6 can be seen passing through the floor before being connected to the pump at the bottom of the well.

The floor boards of the model engine house have been fastened in position using hand forged nails made to represent the flat board nails which would have been used in 1712. Each nail was positioned into a miniature hole of 0.020" diameter.

In order to give the model an authentic appearance, the floor of the basement was covered with small pieces of micaceous sandstone. Each small piece was cut by hand before finally being cemented into position. The choice of micaceous sandstone for the floor was made because it easily splits into very thin pieces which give the appearance of true sandstone which has a very fine texture. There are many dwellings which have their roofs covered with this type of material in North Staffordshire and Cheshire and it is generally referred to as Macclesfield Stone. The stone for the model was obtained from a quarry near Bollington in Cheshire. At this quarry the thinner pieces of stone are discarded, as they have no commercial value, but they are ideal for model building as they have a thickness of only 4mm. The location of this site is at Kerridge Hill and is locally known as Sycamore Quarry.

Reference Number 53 (Fig. 59)
A little Bench with a Bass to rest when they are weary

Over the years, many men have been proud of attending and keeping steam engines working at their best, but the man chosen to attend this engine must have felt very satisfied as he saw water flowing away from the mine workings. When all the adjustments were completed, this is where he sat and rested.

Reference Number 47 (Fig. 60)
The Great Door-Case, 6 Feet square, to bring in the Boyler

By looking at this unique building from the surrounding area, the great overhead beam could be seen moving 'into and out of the house'. Also, the surplus water could be seen flowing away after being lifted from the mine workings. For the first time in mining history, miners could work and extract coal from greater depths than before without the constant fear of flooding.

What was behind this great wooden door was a jealously guarded secret. On no account would this door be opened to uninvited guests! Many people travelled from all parts of the British Isles and also from abroad to see for themselves this great discovery, but they were only permitted to view the engine house from the outside. They could not discover how the steam was used to create the force needed to lift the water from the mine workings. The

Fig. 60: The great door 6 ft square. The brickwork is layered in Flemish bond and notice the 'bullions' in the glass panes of the window.

Spanish ambassador at the English court came with a number of visitors and is said to have departed without discovering the secret.

Thomas Newcomen, after so many years toil, was eager to gain the most he could for himself and his family. He was not going to reveal to others how this mechanical movement was achieved. Marten Triewald, in his 1734 account of the development of this engine, said it had taken ten years of experiments to reach the level of development seen in the engine at Tipton.

Thomas Barney must have taken many measurements in the preparation of his 1719 drawings, because he actually states the size of this door as being 6 ft square, large enough to bring in the boiler. By studying these references and relating their sizes to other components, the miniature engine house could be planned. The door must have been very strong, with large draw bolts firmly holding it closed. A lock was also placed on the outside. The model engine house also features the door, made of Japanese oak, and a lock has been incorporated. This is a carefully made mortice design with a two lever locking mechanism. It actually works and holds the doors firmly closed!

Also shown in the drawing is a wooden window frame with twelve small panes of glass. In 1712, these would have been made by the traditional method of gathering glass which would be blown manually before being formed into small sheets on a level surface. These sheets were then cut to the required size. Making glass by this early method produced window panes of a small size as the glass blower made circular pieces from his blowing iron, because many more small panes could be cut from a circular sheet. Where the blowing iron had been disconnected from the glass sheet, it produced an imperfection which, over the years, has become known as a bullion. This bullion is often incorporated into the frames of modern homes, giving a very pleasing visual effect.

The decision to assemble the engine at Tipton was a wise decision because the glass could have been obtained locally, possibly from Stourbridge, a well known area for glass production.

The windows of the miniature engine house are also made from glass of 0.007" thickness. After heat had been applied to soften the glass, the bullions were made by deforming each piece so that the desired effect of the glass having been made in the 1700s was achieved.

Reference Numbers 54 & 55 (Fig. 61 & 62)
A Man going to Replenish the Fire
The Peck-Ax and Proaker

Illustrated in this drawing is a man about to shovel coal on to the fire beneath the boiler of the engine. In Thomas Barney's drawing of 1719, he is shown in stately dress. This must have been considered a very privileged position. As there is no known portrait or sculpture of Thomas Newcomen, it would be nice to think that Thomas Barney was actually showing the great inventor himself shovelling coal on to the fire!

Also shown are the tools used to tend the fire. The pick is unusual – the coal was delivered in large pieces which had to be broken up before it could be shovelled on to the furnace or fire. The coal came from a seam 30 feet thick producing large lumps which would have to be broken before use.

The figure shown on the miniature engine house is depicted in all his finery and has been made of fine English bone china. He was made by a man who

Fig. 61: The engineer/fireman stoking the fire beneath the boiler.

will be known to all collectors of Royal Doulton figurines, Eric Griffiths, who was the Art Director of the firm. He very kindly sculpted this figure to the scale of the miniature engine. It is the only Royal Doulton figure ever made showing a man about to shovel coal on to a fire beneath a boiler!

Reference Number 56 (Fig. 63)
The Centre or Axis of the Ballanc'd Beam

This is the final reference to Thomas Barney's drawing of 1719 and was described in two distinct parts. The first part of the description was typeset and followed the pattern of all the preceding references, but the following statement had been added in freehand: 'that vibrates 12 times in a minute, and each stroke lifts up 10 gallons of water 51 yards perpendicular.'

Without this handwritten addition, the task performed by the world's first steam engine would have remained unknown because, even with extensive research, by the Newcomen Society in particular, no information has been found which can be added to what is stated in reference 56.

Calculating the Power Output of the Engine

In the early 1780s, James Watt devised a method of calculating the power output of steam engines – this term has been in constant use ever since and is known as 'horsepower'.

Firstly the amount of work the engine performs is worked out. This is a simple calculation and is done by multiplying the weight of water in pounds lifted in one minute by the depth of the mine in feet. This sum is then divided by the constant of 33,000 this is the amount of work a horse could do in foot pounds in one minute.

Fig. 62: The fireman about to replenish the fire, but now he is made from bone china!

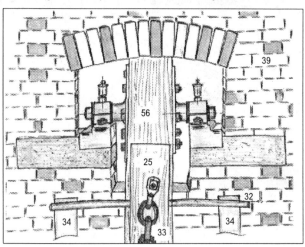

Fig. 63: Details of the central pivot of the balanced beam.

Although Thomas Newcomen never calculated a steam engine's power in this way, it does give an indication of the work done by this engine and, when divided by 33,000, an output of 5.5 hp is the result. However, this can only be an approximate calculation, because in 1712 the Imperial gallon was not always used. Many different standards were in use, the most notable being the wine gallon.

So concludes the description of Thomas Newcomen's engine through the references attached to Thomas Barney's engraving of 1719. All the drawings were taken from the miniature engine.

Fig. 64: The miniature engine and boiler within the sectioned model building.

CHAPTER 8

ENGINES BUILT BY NEWCOMEN'S SUCCESSORS

The world's first steam engine, built in 1712 by Thomas Newcomen, is known to have worked at its original site in the shadow of Dudley Castle for at least twenty-four years. On the 29th October 1736, a cast iron replacement for the engine's original brass cylinder was invoiced to a Darby foundry at Coalbrookdale. The price for the replacement cylinder, complete with a cast iron bottom, was £41-17s-5¼d. The replacement is stated to have been 22" diameter, an increase of 1" over the original brass cylinder fitted in 1712.

At the same time, the engine also appears to have had a complete rebuild. A new boiler was fitted together with the following items: one house water lead cistern, one regulator, one hot well, two cocks, two clacks, a steam pipe, one injection pipe, a sinking pipe, one feedwater pipe to the boiler, a feed pipe to the cylinder, one 'Y' lever, one 'F' lever and two spanners.

Some time before 1740, the engine was removed from its original site to work at the Level Coal Works at Brierley Hill, where it was to work for about twelve years. In c1752, the engine was dismantled and some of the component parts were re-used in an engine at Willingsworth, Wednesbury.

The names of the men who operated Newcomen's engine in the later years is known and they are as follows:

John Hilditch	1725
John Fisher	After September 1731 until December 1736
Edward Fisher	
John Hilditch	In c1736-1737
Daniel Fisher	
Richard Fisher	In c1737-1737 until later than 1740
Edward Fisher	

At this point, we will let the story return to the early eighteenth century. We can now follow the progress of Thomas Newcomen and his successors and see how the steam engine progressed to eventually become the driving force which was used to provide the rotary action to drive the many industries which had previously had to rely on flowing water.

The engines built between 1712 and 1719

This seven-year period will be described next, because during this time twenty Newcomen atmospheric engines are known to have been built. Thomas Newcomen licensed engines to be erected by other engineers and, while describing these early machines up to 1719, it is hoped to say who these engineers were and how they all helped to promote and develop the atmospheric engine into a standard form for mine drainage both in the British Isles and abroad.

Stonier Parrott and the Newcomen Engine

The name of Stonier Parrott, of Audley in Staffordshire and later of Coventry, has long been associated with the early development and the spread of Newcomen's pumping engine, but many of his activities have remained obscure. Considerable information is gathered on his activities by documents in a collection at the Staffordshire Record Office and are known as the 'Aqualate Papers'.

Stonier Parrott was the eldest son of Richard Parrott, of Bignal Hill, Audley. He was a substantial yeoman, very involved in a variety of industrial enterprises at the beginning of the eighteenth century. Richard Parrott's partner was his neighbour, George Sparrow, of Chesterton in Staffordshire, and between them they were 'great farmers, occupiers and undertakers of coals and ironstone and other minerals; of forges and ironworks, and also of brine pits and salt works; within the several counties of Stafford, Warwick, Chester, Worcester and Flint and elsewhere within the Kingdom of England.'

Fig. 65: The oldest portrayal in the world of a steam engine – the Griff Engine. This drawing is entitled: "The engine for raising water (with a power made) by Fire. H Beighton Delin, 1717."

Their involvement included a lease of part of the Griff coalworks of Sir Richard Newdigate of Arbury, Warwickshire. The lease was to run for twenty-nine years and the rent was to be £600 per annum. Soon after taking over the Griff Colliery in 1714, Richard Parrott, Stonier Parrott and George Sparrow entered into an agreement with Thomas Newcomen to erect an engine to draw water from the mine workings, forty-seven yards below ground level – 'An engine to draw water by the impellent force of fire.'

The atmospheric engine made to Thomas Newcomen's design and erected at the Griff Colliery in Warwickshire is shown in Fig. 65. Until 1925, the Barney engraving was considered the oldest portrayal of a steam engine in the world, but in the library of Worcester College, Oxford was found another engraving of a Newcomen engine! This engraving is inscribed 'H Beighton delin., 1717' and illustrates an engine built *after* the Dudley Castle machine; however, the engraving was drawn two years *before* Thomas Barney produced his record of the first Newcomen engine in 1719.

Henry Beighton was closely associated with early Newcomen engines and many writers awarded him a large share of the credit for the development of these early engines. He was born at Griff, near Nuneaton in Warwickshire, in 1686 and spent the greater part of his working life in this area. He died in 1743 and was buried at Chilvers Coton in the parish of Griff.

Beighton was a man of scientific accomplishment who was elected a Fellow of the Royal Society in 1720 and he was the first man of science to make a practical study of the newly invented atmospheric engine.

The generally accepted date for the Griff engine is given as April 1714. The engine was made with a copper boiler, while the powering cylinder and piston were cast in brass. Also stated in the Aqualate Papers was that the pit-barrels and the pipes, together with the cisterns, were cast in iron. This must have been a great innovation because only two years previously elm had been used for these pipes. The engine had a cylinder of 16" in diameter and a stroke of 8 ft. The engine is said to have lifted seventy hogsheads of water forty-seven yards to the surface in each hour. The mine owners, Richard Parrott, Stonier Parrott and George Sparrow, paid a weekly rent of £7.00 to Thomas Newcomen for this privilege.

On seeing the engine pumping water, the mine owners were most anxious to be able to build further engines as there was a requirement for other engines in the Warwickshire coalfields. On the 7th of March 1715/16, they signed an agreement with the proprietors of the patent. Captain Savery had died the previous May and this left Thomas Newcomen, John Meres of London, gentleman; Edward Elliott of London, gentleman; and Thomas Beake of the City of Westminster, tallow chandlers, as the sole 'proprietors of an invention for raising water, and occasioning motions to all sorts of Millwork, by the impellent force of fire.' – this is the wording from Captain Savery's patent taken out on the 25th July 1698.

With the signing of this agreement, the owners of the Griff colliery were now able to construct Newcomen pumping engines for themselves and in the succeeding years numerous engines were built at their mines. The partnership of Stonier Parrott and George Sparrow became widely known and they constructed many of the engines used in the early eighteenth century. Details of the Griff engine are shown in Figs. 66 and 67 overleaf. Extracts from the Newcomen Society Transactions vol. 50, pages 49 and 50, give more details:

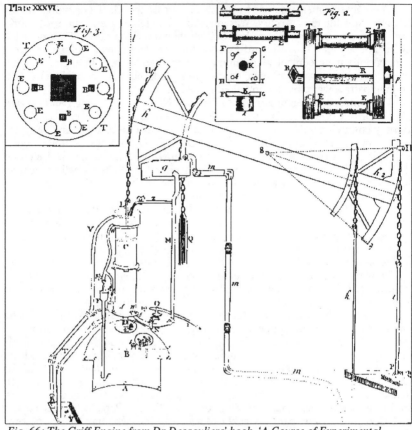

Fig. 66: The Griff Engine from Dr Desaguliers' book, 'A Course of Experimental Philosophy'. Note his other mechanical details on general engineering.

'The profits the Proprietors of the Patent expected to gain from licenses to erect engines were considerable. Richard and Stonier Parrott and George Sparrow had to agree to pay two rents, one of £420 and one of £840 a year, the quarterly payments on both to be made to John Meres at the Apothecaries Hall, London. The Parrotts and Sparrow had leased other mines in Warwickshire at Wood Sydnall, Pickards Fackley, Boys Waste and Lapworth in the parishes of Foleshill, Exhall and Hawkesbury.

'The Proprietors of the patent agreed that they could erect engines as necessary, not only at Griff, but also at the other Warwickshire mines.

'For their part the Staffordshire partners agreed to keep the engines in repair and to allow the Proprietors to repossess for themselves the first Griff engine at the termination of the lease.

'This is how the partnership of Stonier Parrott and George Sparrow became associated with the Newcomen Engine. They now hoped to capture the market with this new invention and, between 1714 and 1732, they were concerned in varying degrees with the erection of at least fourteen engines.'

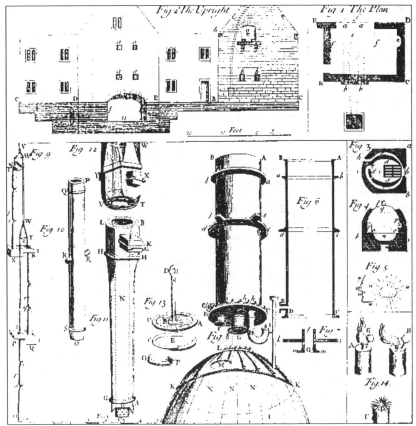

Fig. 67: More Griff Engine details, showing the two stage pumping arrangement using pump trees.

The Third Engine

This was an engine built at Woods Mine, Hawarden, in Flintshire. The exact date of construction is not known, but the engine was built between April 1714 and before November 1715 and it was the first to be built in Wales. The colliery was owned by Thomas and Richard Beech of Stone, Staffordshire, and William Probert of Hamner, Flintshire. The owners appointed Stonier Parrott as their salaried agent at the coal works. Thomas Newcomen, ironmonger, held the other third of the colliery by deed poll.

The building of this engine is usually attributed to Thomas Newcomen, but all the partners appear to have been involved – perhaps it was a combined effort between the engine's inventor and Stonier Parrott. The engine was completed by July 1716 at a stated cost of approximately one thousand pounds for the complete erection and fitting up. The task performed by this engine, and its cylinder size, are not known.

The Fourth Engine

John Calley, Thomas Newcomen's partner, is attributed as the builder of this engine at Moor Hall, Austhorpe, Leeds in 1714/15. This engine had a short working life of four years and raised water from the mine workings thirty-seven yards below ground level.

The cylinder had a diameter of 23" and a working stroke of 6 ft. The engine appears to have had two methods of operation, because it is recorded that it made fifteen working strokes per minute when it was manually operated, but when working automatically it produced only twelve working strokes per minute.

The boiler which produced steam for this engine was set into the brickwork 2 ft 8" above the fire-grate, and was said to burn 24 to 25 'corves' of coal in twenty-four hours. Soon after commissioning this engine, John Calley died at Austhorpe, at the comparatively young age of fifty four.

Calculating the Power of a Newcomen Engine

Many mine owners must have been wondering just how powerful these atmospheric engines were? To answer this question, Thomas Newcomen devised the following method to calculate the power output of the pioneering atmospheric engines and stated their power in 'long hundred weights'.

Newcomen squared the diameter of the cylinder in inches and, after dropping the last figure, he then multiplied this by a known percentage of the atmospheric pressure and called the result 'long hundred weights'. He then wrote a cipher on the right side and called the number on that side 'odd pounds'. This, he considered, was accurate when the atmospheric pressure was high, or when the barometer was reading 30" of mercury.

The effect of removing the last figure from the square of the diameter is equivalent to dividing the area of the cylinder into portions of ten circular inches. The force exerted on each of these portions is estimated as a long hundred weight. It was called such because the result was 120 lbs. This figure was then divided by ten, which gave the force which would theoretically be exerted on each circular inch. To compensate for the engine's inefficient performance, Newcomen then allowed between one-third and one quarter part of the earth's atmospheric pressure as his standard.

Reading through this, it is little wonder that a steam engine's output was eventually measured in horsepower – a unit which can easily be visualized and understood by all who wanted to know how much work an engine could perform.

Stone Pitts, Ginns, Whitehaven (Engine No. 5)

Of all the twenty atmospheric engines which are thought to have been built between the years 1712 and 1719, an engine built in 1715 by the combined effort of Thomas Newcomen and John Calley is the best documented. The record shows that it had a very long working life because in 1736/37 the power output was increased by fitting a larger powering cylinder of 38" in diameter.

The original specification of this engine stated that a cylinder of 16" in diameter and 8 ft in length should be fitted, but the cylinder which was eventually fitted had a diameter of 17" and was cast in brass.

The mine owner, Mr James Lowther, agreed to pay the proprietors of the patent a yearly rent of £182 on completion of this engine.

James Lowther's mining agents were the brothers John and Carlisle Spedding. Letters exchanged between these associates tell the story of many abandoned mines by 1705. This was due to the old water gins, together with the wind driven pumps, being unable to cope with the influx of water at these greater depths.

The mine owner and his agents studied Captain Savery's method of drainage and concluded that their requirement was far greater than the pumping capacity of his 'miner's friend'.

The newly developed atmospheric engine seemed the only course to pursue. In August 1715, James Lowther held discussions with Thomas Newcomen who was asked to submit his proposal to drain the flooded mine at Whitehaven.

Newcomen's proposal is said to have been well presented with great accuracy and minute attention to detail. After the agreement was signed, a drawing of the engine house arrived and the final preparations went ahead.

The boiler was made in Newcastle and brought to Whitehaven on wagons which had to be widened to carry the large load. Many of the other engine parts arrived by ship from Bristol and Dublin, as there was no direct service with Whitehaven at this time. The engine was built and tested during the winter of 1716/17 and, after many teething troubles, mostly with the pump trees and boiler corrosion, the engine is said to have worked well, lifting 140 hogsheads of water to the surface in one hour. It was not a vertical lift from the flooded mine – the first twenty-three yards down were perpendicular, then the next thirty-two yards were negotiated by a sloping shaft.

The boiler of the engine was said to have been made of iron and fitted with a dome of beaten lead. By April 1717, difficulties had arisen with the boiler because of corrosion. Newcomen recommended that a new boiler should be made and this eventually arrived by sea from Frodsham. The boiler was supplied by Stonier Parrott from Bignall Hill in Staffordshire and again had a dome made from beaten lead.

The original 17" diameter cylinder was increased to 22" in 1727, and in 1736/37 a large cylinder was fitted, with a diameter of thirty-eight inches. This was a large cylinder for 1736 and would most probably have been cast in iron.

Many atmospheric pumping engine installations of about 1736 used cast iron for their powering cylinders as it was a cheaper material than brass, while at about this time machinery was becoming available which could bore the internal surface of the cylinders, providing a better seal for the moving piston. The machinery developed for cannon boring was used to machine the early cast iron cylinders. The one great advantage which cast iron had over brass was its lower thermal conductivity. This would have made the engines more efficient, as less heat would now be wasted at each stroke of the engine.

None of the original engine house walls now remain, but the faint outline of where the building once stood can be still be found on the site.

Yatestoop Mine (Winster), Derbyshire Engine Number 6

Stonier Parrott's partner, George Sparrow, is attributed with building the first engine to be used outside the South Staffordshire coalfields. This engine was used to draw water from a lead mine near Matlock in Derbyshire.

In May 1716, George Sparrow agreed with the proprietors of the lead mine at Yatestoop, Winster, to erect 'one or two engines according to the present

Fig. 68: Another drawing of an engine from Desaguliers' 1744 book. This engine is thought to be the Oxclose engine.

acceptation of the term fire engine' to draw water from the mine workings. Sparrow agreed to erect an engine at this mine within a twelve month period and, in return, he would receive from the mine owners a $\frac{1}{7}$ part of all the lead ore mined from below the flooded water level of this mine.

The only details to survive about the engine are that the vertical powering cylinder was made of brass and that the installation was completed by July 1717. However, George Sparrow must have had great success because this engine was the first of three to be built by him at this lead mine.

Engine Number Seven

This engine was built at Washington Fell, County Durham, and has become known as the 'Oxclose Engine'.

The illustration shown as Fig. 68 is drawn with great precision and is generally thought to show a mirror image of the Griff engine engraving. It is attributed to Beighton and is believed to show the engine which was built at Oxclose, Washington Fell. The exact date of construction of the engine is not known, but it was completed before December 1717. In the same year as he made his engraving, Beighton produced calculated data in tabulated form which would enable mine owners and engineers to determine the amount of water an engine would pump from a given depth.

There are no details surviving on the Durham engine, but the tabulated data is based upon an engine working at sixteen strokes per minute, with a working stroke of 6 ft and Beighton probably based his calculations on this northern engine. The tabulated information was not published until 1721 and then not in the usual scientific journals, but in *The Ladies Diary*! Beighton must have been a very versatile man, because he even found time to be the editor of this publication.

Fig. 69: Steam control valve mechanism thought to have been fitted generally to steam engines in about 1716. Note the improvements from the 1712 Dudley Castle engine.

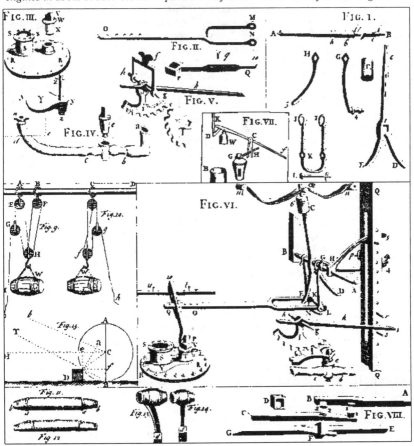

91

Byker Engine, Newcastle-on-Tyne

The building of atmospheric pumping engines in the early part of the eighteenth century must have been the sole occupation of all who possessed any knowledge of their installation. In 1717 an agreement was signed to install a large engine at the premises of Messrs Ridley, of Newcastle-on-Tyne, who had a mine rich in coal but no successful means of removing the flooding water. The only person not fully occupied at this time was thought to be John Calley's son, Samuel. Although only sixteen years of age, his father thought he knew enough to supervise the installation of this engine.

Marten Triewald, the Swedish engineer and philosopher who had worked with the original inventors of the atmospheric engine, Thomas Newcomen and John Calley, thought the ideal opportunity had arisen for Samuel Calley to learn about the details of these engines. This he did by offering to assist him with the building and commissioning of the engine at Newcastle. The mine owner, Nicholas Ridley, was very pleased with this arrangement and, after some detail changes suggested by Triewald, the engine was built and soon drew water from the mine workings for an annual rent of £420.

Never one to show great modesty, Marten Triewald actually said: 'Mr Nicholas Ridley, who was then staying in London, was not only very much perturbed because of the youthfulness of his engineer, but also feared that his competitors and other owners of coal mines in the neighbourhood would get an opportunity to corrupt this youth, so that he would not serve him faithfully. In these precarious circumstances Mr Ridley, who had known me from early childhood, and moreover was aware with what diligence and zest I had been studying natural science and mechanics in London, was led by the wonderful foresight of God to the idea of persuading me to assist his young engineer, and to watch him in case he did not serve his master with all the faithfulness and honesty required in such a delicate position. Thus he persuaded me to go with him to New-Castle, promising to promote me to the knowledge of how to construct fire-machines, and I for my part promised to serve him loyally against a fair reward.

'A few days later we arrived in New-Castle where the construction of the first fire-machine in this district was in full swing, I for my part, according to our agreement, did not allow anybody within the space of a year and a day to discover in the slightest way that I understood anything whatsoever about such a machine, though I, however, as soon as I saw the machine at work, conceived a more complete theory of it than the inventors themselves possessed down to the very moment of their death.

'Even the great Newton testified in the presence of the Commission-Secretary Herr Skutenhielm that such was the case when he admitted that he never had been able to get a correct idea of the fire-machine from the inventors, because they always ascribed the power to the steam, which, however, only constitutes the agent by means of which the power is obtained.'

These are two extracts from Marten Triewald's book, first written in 1734 and translated from the Swedish into English by Mr Are Waerland in 1928. Marten Triewald altered Thomas Newcomen's original specification of this engine by increasing the cylinder to 33" in diameter from the original specification of 28" diameter. The engine had a working stroke of 9 ft.

After successfully building this engine, Triewald and the young Samuel Calley formed a company to supervise the construction of 'fire-machines'.

They divided the income from this partnership equally between themselves. Triewald must have shared John Calley's confidence in his son's ability at such an early age.

Between the years 1717 and 1719

This must have been a furious time of industry because no fewer than thirteen engines were constructed in this period. The partnership of Stonier Parrott and George Sparrow was responsible for all but one installation. However, far less technical interest was now being shown in the engines because on no installation has any record survived stating the cylinder size or the stroke of the engines. On only one installation, Park Colliery, Tyneside, is it stated how much water was brought from the mine workings to the surface.

By 1719 the atmospheric pumping engine must have been an accepted part of coal mining and the engines were used wherever water was present in the mines.

CHAPTER 9

THE OLDEST SURVIVING ENGINE HOUSE

The model of Thomas Newcomen's engine was completed by October 1987 and, after spending so much time in the research and construction, I was very pleased to be able to display this engine for the first time at the 1987 Midlands Model Engineering Exhibition, where it was awarded first prize and also the Modelcraft Cup. After this initial showing, the model was taken to the model engineering exhibition held in London. The model was again successful, gaining the first prize and a gold medal. From this exhibition, the model attracted the attention of the presenters of BBC Television's 'Blue Peter' programme, with the result that the model was shown working on the show on 14th January 1988. Great interest was shown nationwide, but the correspondence which proved the most interesting was from Mr James Boyle of Ardrossan, Ayrshire.

Mr Boyle is a member of his local Museum Association, who asked if he could have some more details of the model. He asked if I could supply photographs and a comprehensive description of the original engine because he was organising, at his Museum, a display of colliery machinery centred around a large illustration of a well-known local landmark, a ruined building which once housed a mine pumping engine thought to have been built in 1719 to draw water from the mine workings of a colliery at Stevenston, Saltcoats, Ayrshire. The photograph, Fig. 70, shows how the early engine house appeared about 1938.

It was not until these early engines were being described in detail and in chronological order that the date of 1719 became of some importance. If this could be proven to be the age of the ruined building, then we are looking at the oldest surviving relic of a Newcomen engine house in the world!

At this stage, the Newcomen Society were asked for their opinion of this relic and, on seeing the photograph, gave credence to this being a very early atmospheric engine house indeed. But could the date of 1719 be proved? The only written evidence to be found was from a report entitled *Excerpt from Burgh of Saltcoats Quarter Centenary 1528-1928*, which states the engine house was erected in 1719 by the Saltcoats Shipmasters, lessees of the colliery and saltpans.

The engine was installed for the pumping of water out of the pits and was stated to be the second engine set up in Scotland. The report further states that the site of the engine house was close to the course of the canal which was built to transport the coal from the colliery to Saltcoats harbour for shipment. The canal was completed on 19th September 1772.

At this juncture, Historic Scotland was contacted and John Hume was willing to try and authenticate the date of 1719. John has written on industrial archaeology in Scotland and seemed very interested in this project. He even said that Historic Scotland was willing to carry out an archaeological dig if

Fig. 70: The Scottish engine house photographed circa 1938. By kind permission of Saltcoats Museum.

documented evidence was not forthcoming. Unfortunately, no new evidence has been found, but Historic Scotland is quietly confident that the date of 1719 is the correct one and this ruin was originally built to house the first Newcomen engine ever assembled in Scotland. The second engine to be built in Scotland was at Elphinston Pit, Tranent, East Lothian in 1720.

What is Known?

The only reference to an engine assembled at the Stevenston Colliery in 1719 is in *The Steam Engine of Thomas Newcomen* by Rolt and Allen, where they mention that the engine was originally built by Peter Walker and John Potter and had a powering cylinder of 18" in diameter. This was cast in iron and is generally thought to be the twentieth engine built after Newcomen's original Dudley Castle engine of 1712.

The Engine House Today (1996)

The engine house remains are situated on what is now Auchenharvie Golf Course, easily found between Stevenston and Saltcoats. Many of the detailed features are now lost from the historic photograph, Fig. 70 above, as the building has been consolidated to prevent any further decay. Sadly, it is not now easily recognised as the building that once housed such an important engine.

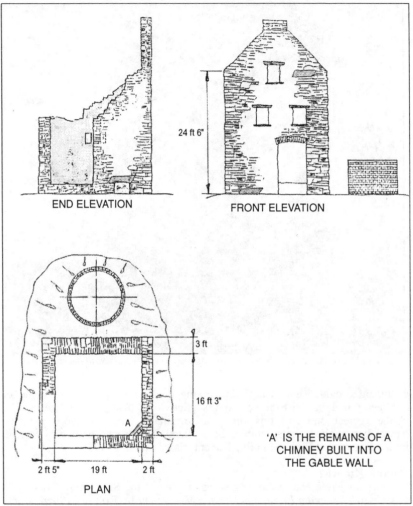

END ELEVATION

FRONT ELEVATION

24 ft 6"

3 ft

16 ft 3"

A

2 ft 5"   19 ft   2 ft

PLAN

'A' IS THE REMAINS OF A
CHIMNEY BUILT INTO
THE GABLE WALL

Fig. 71: A drawing showing some of the dimensions of the engine house.

On a recent visit to photograph and measure the engine house, I was often asked why anyone should be measuring the remains of this building. The local children thought it was the remains of a Scottish castle! The standing walls are built from random sandstone, interspersed with large blocks of granite.

After sketching and taking most of the main dimensions, the conclusion was reached that this relic had originally housed an atmospheric pumping engine, the dimensions indicating that it was built to hold a very small engine. The dimensions showed that the whole installation was much smaller than the engine at Tipton and a drawing showing the major dimensions of the building is illustrated above.

Fig. 72: The consolidated remains of the engine house photographed in 1996.

A close study of the original photograph, together with a detailed study of the building's features, reveals this to be a most unusual installation, because when this engine was working, the great overhead rocking beam did *not* pivot within the end gable wall but passed through and pivoted on one of the adjoining side walls. The evidence to substantiate this theory is that the wall on the left-hand side of the photograph is almost 12" thicker than the three remaining walls. The accepted practice was for the great overhead rocking beam to be supported, and pivoted through, one of the end gable walls, but on this early atmospheric engine house, the standing gable wall was the narrowest of the whole building! Conclusive evidence that this was so can be seen in

97

Fig. 70 showing the round structure of the original well head adjacent to the side wall.

Proof that the beam was supported on the side wall can also be seen by two window-like openings. In the early eighteenth century, these openings supported the great sommer beams which held the powering cylinder in the vertical working position and, by some careful measurements, they reveal that these beams were about 20" apart. This measurement would allow the 18" diameter cylinder to be held firmly in the vertical position with long nuts and bolts. This would suggest that the structure was the original building of 1719 and not a later construction, as the early eighteenth century pumping installations were only superseded by engines using larger cylinders which was considered at that period the only way of increasing the engine's power, so the Scottish building could not have accommodated an engine with a larger cylinder!

From the written facts and dimensional evidence, it appears that the building on the Auchenharvie Golf Course held the first Newcomen atmospheric engine in Scotland. It is hoped that the remains of this building have continued protection and a plaque stating it to be the oldest surviving structure to have ever contained a steam engine will be fitted.

In Sweden, the building which once housed the first Newcomen engine constructed by Marten Triewald in 1728 still survives at Dannemora. It was considered to be the oldest building to have contained an atmospheric pumping engine, but now all the available evidence suggests that the Scottish relic predates the Dannemora engine by nine years, so the Scottish structure is by far the oldest surviving building to have housed an atmospheric engine.

CHAPTER 10

WATERWORKS ENGINES

An atmospheric Newcomen engine built in 1726 started an unbroken succession of engines which extended well into the twentieth century. This engine was the first of the domestic waterworks pumping engines and was built on the banks of the River Thames to pump water for distribution throughout London.

The London Waterworks, known as the York Buildings Waterworks, was established in 1676 and was situated at the lower end of Villiers Street, Strand. Captain Savery installed one of his engines at these waterworks, but without great success. When the Newcomen engine was installed, it stood alongside the Savery engine. The engine had a large brass cylinder of 35$\frac{1}{2}$" diameter.

The engine, Fig. 73 below, was drawn in 1725 by Sutton Nicholls and is thought to be the engine assembled at the York Waterworks. Prints of this well documented engraving were offered for sale at a price of one shilling

Fig. 73: An atmospheric engine house from an engraving made in 1725 by Sutton Nicholls.

Fig. 74: A Newcomen waterworks pumping engine. The construction is much lighter and smaller than the engine shown in Fig. 73.

by John King at The Globe in the Poultry, London.

The stately dressed gentleman viewing the country scene had been shown on all the illustrations of Newcomen engines until this date. He was shown for the first time on the drawing of the Griff engine which was drawn by Henry Beighton some eight years before. Beighton set a standard way of illustrating these early Newcomen engines, with the figure being used to give a sense of scale.

The only difference between many of the early steam engine pumping installations was the physical size of the engine which determined its ability to lift water from mine workings or for domestic use. Another illustration of a similar, though much smaller, Newcomen waterworks engine is shown as Fig. 74.

From the original Newcomen waterworks engine built in 1726, there evolved some of the largest and most powerful steam engines ever to be made. The engines were often floridly and colourfully decorated with much attention to aesthetic detail, the builders even applying gold leaf to many of the structural supports of the building and engine castings. In many counties, the pride in the municipal waterworks showed itself in the pumping engines' ornate buildings, which have been described as 'cathedrals of steam power.' Only a few of these superb buildings are still standing today. One of these magnificent steam pumping stations was built by the Staffordshire Potteries Water Board at Hatton in 1890 and is shown in Fig. 75. Engines continued to be built until the early 1920s, being powered by steam at very high pressure and were used until the early 1960s, only succumbing to high powered submersible electric pumps which were much cheaper to run.

Fig. 75: A pumping station built by the Staffordshire Potteries Water Board at Hatton in 1890.

CHAPTER 11

ENGINEERS FROM OVERSEAS

The first Newcomen engine to be built outside Britain

The first atmospheric engine to be designed and built outside Britain was built in 1728 in Sweden. The man responsible for this engine was Marten Triewald, who took Newcomen's ideas and built an engine to pump water from the mineral mines at Dannemora near to Stockholm.

This is a well-documented engine, with many of the original drawings still surviving in a Swedish archive, but before studying this engine in detail, let us look at the man responsible.

Who was Marten Triewald?

Marten Triewald was born in 1691 and was the son of a smith whose business must have prospered, because he was able to give Marten a good education.

Fig. 76: A portrait of Marten Triewald.

At the end of Marten's schooldays he initially started in commerce, but this proved to be most unsatisfactory and it was in 1716, at the age of twenty-five, that he came to England, initially to study the technological developments of the early eighteenth century.

In London, he became acquainted with a good circle of friends, including Dr Desaguliers. It is thought that Triewald attended Desaguliers' lectures on natural philosophy and also met Nicholas Ridley, from whose association he learned about the operation of the newly developed atmospheric engine.

Marten Triewald stayed in England for ten years and his time was devoted to working with Thomas Newcomen's newly developed atmospheric engines and lecturing on physics and mechanics. He wrote about a great variety of subjects, such as the silk industry, the manufacture of soap, the cultivation of hops and bee-keeping.

His collection of scientific instruments is a treasured possession of the University of Lund. It is one of the scientific distinctions of Triewald that he, conjointly with the great botanist Carl Linnaeus, was the founder of the Royal Academy of Science in Stockholm in 1739. He had also the added distinction, of being made a Fellow of the Royal Society of London.

In 1734 he wrote *A Short Description of the Atmospheric Engine*, a compilation of records of the time he spent in England working on the early atmospheric engines, culminating in his own account of the Dannemora engine.

Marten Triewald died in 1747 at the comparatively early age of fifty-six. There are only three known copies of Triewald's book in England – one at the Royal Society, another at the British Museum and the third at the Patent Office Library. The copy from the Patent Office was used to translate this important work into English for the Newcomen Society in 1928.

The Dannemora Engine

The description by Marten Triewald of the engine was written in 1734 and closely follows the ideas set out by Thomas Barney in 1719 with his notes attached to the Dudley Castle engine drawing.

Triewald also produced a drawing of the Dannemora engine which is shown as Fig. 77. On the drawing are reference numbers and letters from A to Z, although not many of these references are easily seen. They again follow the style set by Thomas Barney.

Triewald even starts his description of the engine with the fire place, followed by the boiler and then the powering cylinder. Working with the steam engineers in England between the years of 1716 and 1726 must have had a profound influence, because not only did the Dannemora engine look like the English counterparts, but it was recorded in the same manner!

The fire door to the furnace of this engine must have been bigger than its English counterparts; a close study of Fig. 77 overleaf will reveal that this engine was a wood burner. Neat lengths of wood can be seen stacked outside the engine house, ready for use. The man standing by the boiler dome seems to be passing the wood through a trap door to be loaded on to the fire below.

The boiler of the Dannemora engine was of large proportions and was made in Stockholm by a coppersmith. Constructing this boiler 11 ft in diameter out of beaten copper must have been quite an innovation in 1727, because at this period lead was the material usually chosen for making boiler domes; only the boiler base was usually made from copper.

Fig. 77: The 'Dannemora Engine' of 1728, the first engine to be built outside Britain.

The cylinder which was used by Triewald to power the engine had a diameter greater than any which had been used before. Marten Triewald tried unsuccessfully to obtain this cylinder in England. The 36" bore diameter cylinder, with a designed stroke of 9 ft, was cast in Stockholm by Gerhard Meyer whose family managed the Royal Gun Foundry. This large component was cast in gunmetal, which consisted of 90 parts copper alloyed with 10 equal parts of tin. Cast upon the cylinder were the names of Marten Triewald and Gerhard Meyer with the date of 1728. This was surrounded by a decorative embellishment. Triewald is said to have been very pleased with the result, as the cylinder had a very smooth bore and the piston was an excellent fit.

It is not intended to describe all the references to the Dannemora engine but to select items of greatest interest.

The boiler appears to be fitted with a new innovation, because it is stated that a valve was used loaded with deadweights, which could be varied. This was to allow the steam to escape if it 'gets too strong'. This appears to be the first reference to a boiler fitted with a safety valve.

The piston was a disc which closely fitted the internal diameter of the cylinder. Around this disc was fitted a thick layer of felt and, to achieve a complete seal between these two items, water to a depth of 6" always had to be present. When the seal was complete, the engine was said to exert a force of 21,575 skalpund, or 53 skeppund and 18 lispunds. This was the force the engine was said to exert on the chains attached to the great overhead rocking beam.

The method used by Marten Triewald to measure an engine's output will be compared with that used by Thomas Newcomen at the end of this section.

One unusual feature incorporated into the copper boiler of the Dannemora engine was the fitting of a wooden lid to the inspection hatch. This opening into the boiler must have been quite large, because it is stated that it was big enough to get through and into the boiler when necessary for descaling and routine maintenance work. None of the access doors on the early English engines were known to have been made of wood.

Before the building of the Dannemora engine in 1727, never had the great beam of an atmospheric engine been designed to transmit such a high force. This engine, with a cylinder 36" in diameter, would transmit a force of 15,964 pounds (this is a theoretical force with a maximum vacuum) by means of a 4" thick tension chain to the wooden rocking beam positioned above. This great beam measured 30 ft long and was made from six baulks of pine, firmly held together with wrought iron straps and bolts. The composite beam shown in Fig. 77 is an innovation because the installations before this date are shown to be made from single pieces of wood, usually oak.

The arch heads, or quadrants as Triewald referred to them, were made of oak.

The pump forcing the water to the surface from the mine workings was also of a composite nature, because the pump body was stated to have been made of oak and fitted with an 8" diameter metal plunger.

The description by Marten Triewald of this Swedish engine is written with great enthusiasm, and he extols the engine's many theoretical virtues. However, the mine owners did not share Triewald's delight with this engine and it was not a great success. All the promises to keep the mine free from water were not forthcoming, and the mine owners reverted to the original method of drainage using horse powered whims. Could this engine's lack of success be attributed to the Triewald method of calculating the power output? He calculated the force exerted by an atmospheric engine in a completely different method to that used by Thomas Newcomen.

The unit of force used by Marten Triewald to measure this engine's power in 1727 has long become redundant. In Sweden at this time, the weights were measured in Skalpunds, Skeppunds and Lispund, and the volume of water removed was gauged in Kannors, which is approximately equivalent to half an Imperial gallon. Triewald made his calculations difficult to understand, as he used a combination of these early Swedish units and the English Imperial measurements, as can be read from the following extract taken from his book:

'A pillar of water 34 feet high and of the same base as the area of the piston of the Dannemora fire-machine, ie. 1018.3 square inches, weighs 21,575 skalpund and 53 skeppund 18 lispund 15 skalpund victualie weight, which therefore constitutes the power of the machine.'

One skalpund equals 13.7 troy ounces, one skeppund equals 374 Imperial pounds, and one lispund equals 18 Imperial pounds 12 ounces. On reading through his written account of the power output of this engine, and the associated calculations, it leaves little doubt that his claims for drawing such large quantities of water from the mine workings two hundred feet below were exaggerated. If Triewald had sold the idea of mine drainage to the owners based upon these calculations, it is little wonder that they reverted to the use of horse whims! He actually states, that his one engine would replace and draw as much water in twenty-four hours as sixty-six horse whims could, each using four pairs of horses.

Unlike Thomas Newcomen, Marten Triewald took no account of his engine working on any less than the maximum theoretical atmospheric pressure. The column of water he talks about would exert a force of 14.7 pounds per square inch upon the top of the piston. His calculations are then based upon the formation of a perfect vacuum to draw the piston to the other end of the cylinder.

The method used by Newcomen, when calculating the amount of work an engine could do, was to take *two-thirds* of the actual atmospheric pressure; this was the vacuum he could re-create on the underside of the piston. This method of power calculation satisfied all the installations in the British Isles throughout the eighteenth century.

The power output of the Swedish engine is stated as 21,575 skalpunds. This force, when converted into Imperial units, would have been impossible to achieve. Trials were conducted over the following three years to try and perfect this engine, but they only produced a limited success and after this time the owners of the mine reverted to the original drainage method. The engine remained in its original building until c1775, but all that now remains is the engine house as shown in Fig. 77 on its original site but now empty of any fittings.

Jean Theophile Desaguliers Li.D. F.R.S.

The one person who is inextricably linked with the documentation of early atmospheric engines is the Reverend Jean Theophile Desaguliers (1683-1744). In 1744, Desaguliers published a work which was to become a classic document for early engine research, entitled *A Course in Experimental*

Fig. 78: The Reverend Jean Theophile Desaguliers Li.D., F.R.S.

Philosophy. The tome contains many engine drawings showing the most intricate of details which, when put together, would make an early atmospheric engine.

Desaguliers came to London in 1713 and quickly established himself by befriending the scientific thinkers of the day. In 1716 he exchanged ideas with Marten Triewald and neither of these men were able to disguise their jealous thoughts towards Thomas Newcomen's achievements.

Desaguliers states in his book that if he had thought of creating a vacuum by the condensation of steam before Thomas Newcomen, he would have had a better understanding of the principles involved, and would have made a better fire powered engine. He did not give Thomas Newcomen any credit for his discovery, agreeing with Marten Triewald that it was most likely the result of an accident.

Desaguliers was made a Fellow of the Royal Society and is said to have been an impressive teacher and lecturer who became the Society's curator of experiments. He was chaplin to His Royal Highness Frederick, Prince of Wales, and also to the wealthy James Brydges (Earl of Carnarvon) at whose house, Cannons, near Edgware, he designed and installed the domestic water supply. He was also Grand Master of the Freemasons.

Desaguliers' contribution to steam engineering may have been modest, but he has had a great influence through his writings. He made many attempts to perfect the Savery principle of drawing water by the combined forces of a vacuum and the expansive power of steam. His attempts were mainly designed for the wealthy, who used the Savery machine to pump water for their garden fountains. A drawing of one such installation is shown in Fig. 79.

Fig. 79: Desaguliers' improved Savery engine (1718) designed to ornament garden fountains.

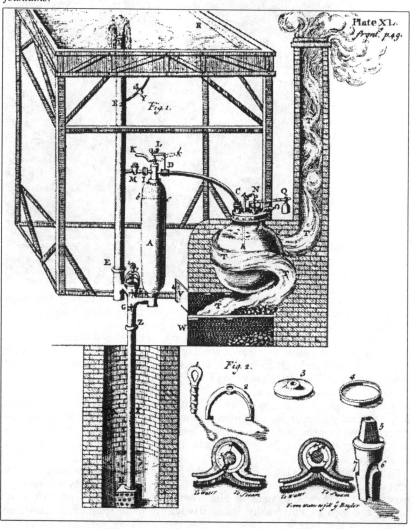

In 1718 Desaguliers befriended Henry Beighton, who had been responsible for the engraving of the Griff engine built in 1714. The two men had a good understanding of scientific principles and worked well together.

Fig. 68, which is illustrated in Chapter 10, is from Desaguliers' *A Course in Experimental Philosophy*, but a close study will reveal that it is actually a tracing from Beighton's earlier drawing of the Griff engine. The drawing in Desaguliers' book is of an engine thought to have been assembled on the Oxclose site. By tracing the earlier drawing, the artist has reversed the image. The only dimensional difference between the two drawings is that the gable walls on the later drawing are two feet further apart. There has also been a modification to the principle of the valve operating mechanism on this drawing. All the other dimensions of the drawing are identical; they even have

Fig. 80: Two pumps redrawn from Desaguliers' book, 'A Course in Experimental Philosophy'.

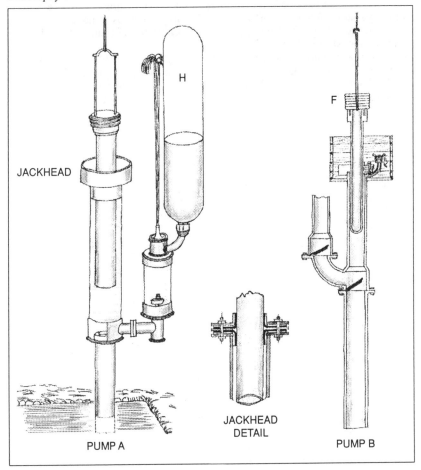

the same number of layers of bricks in each gable wall. The artist producing this second illustration must have thought the country scene was too tedious to reproduce and did not include it in Desaguliers' book.

Most of the drawings showing atmospheric engines up to about the year 1725 seem to be a variation on Henry Beighton's drawings of the Griff engine. They all possess the well-dressed gentleman viewing the country scene beyond the pump rods as they descend into the mine workings below.

Illustrated in his book, Desaguilers shows the typical type of plunger pump which was used with these early atmospheric pumping engines. Fig. 80 shows two pumps redrawn from Desaguliers' book. Pump A was used for ornamental fountains where the suction lift was usually small, the only requirements being a very pleasing visual column of water. An accumulator, marked H, was stated as an 'Air Vessel to joyn to any Conduct pipe after the Forcing Valve.'

Pump B was a plunger pump invented by Sir Samuel Morland (1625-1695) and was the type used for mine drainage by the early pioneers. A reference to pump B states, 'A section of the Forcing Pumps such as is used often in the Engine to raise Water by Fire to show the working of the Plunger thro the Jack head and the Motion of the Valves.'

This type of pump is still in everyday use pumping clay slip in the pottery industry. The use of the weights F to vary the pumping pressure gave these Morland devices a name which has endured to the present day, still being known as a 'deadweight pump'.

CHAPTER 12

THOMAS NEWCOMEN – THE FINAL TRIBUTE

The Year of 1729

Thomas Newcomen died on the 5th August 1729 at six o'clock at the house of Edward Wallin. During his final illness, he had the advice of two 'Skilful Pysitians' every day, and a nurse continually with him, and he is said to have departed this life 'without a sigh or groan'. His death was noted in the *Monthly Chronicle* for August which said, 'died Mr Thomas Newcomen, sole inventor of that surprising machine for raising water by fire.'

He was buried on the 8th August in the Nonconformist Burial Ground at Burnhill Fields, Finsbury, London. The register records, 'Mr Newcomen from St. Mary Magdalens, buried in a vault, 00-14-00', but sadly, the position of this vault is not known and no headstone is thought to have been erected. His wife, Hannah, lived for a further twenty-seven years and died in 1756.

Fig. 81: Upper Street, Dartmouth, in 1839. Thomas Newcomen's house is in the left foreground. From a watercolour by Miss C B Hunt.

A watercolour painted in 1839 by Miss C B Hunt shows where Thomas Newcomen was living at Dartmouth when he invented the atmospheric engine but, sadly, this house was demolished in 1864. No finer tribute can be paid to Newcomen's achievements than that written by L T C Rolt in 1963: 'With little capital, no machine tools and no text books to help him, Newcomen, captaining a team of craftsmen on the site, succeeded in building a machine so masterly in design that, in its broad essentials, it endured for nearly 200 years. It was a feat without parallel. His was a truly archetypal invention, so sound in principle that, once conceived, it formed an indestructible foundation upon which posterity could confidently build. By showing the world how power could be harnessed by means of a cylinder and piston, Thomas Newcomen pointed the way forward which mankind has followed from that day to this with the astounding results we now see all around us.'

Only two letters survive that are known to have been written by Thomas Newcomen. The first was written on the 7th May 1725 from Dartmouth and was to the Lord Chief Justice. The other letter was to his wife and was written on the 30th December 1727. This is a revealing insight into the mind of this God-fearing man.

London December 30th, 1727

My Dear Wife
      I rejoice to hear by yours 26th inst. that the family is in good Health which Mercy I am also favoured with. I suppose Elias may be returned before this comes to Hand, if not remember me kindly to him, and to our other two Children, and tell them I should greatly rejoice to hear they were seriously enquiring the way to Sion with their faces thitherward. This ought to be their Chiefest Concern, as ever they propose to themselves the Enjoyment of True Happiness. Tell them that I sometimes reflect upon the Mellancholy Circumstance of the Late Prince Menzikoff, who a few months since was Prime Minister to the Great Emperour of Russia, had arrived to an Extraordinary Height of Power, had accumulated to himself an Immense Quantity of Riches, and was almost adored by all as the most Happy of Men, but was suddenly deprived of all and reduced to his former Degree of Meaness having also incurred the highest Displeasure of the Great Monarch; yet, in my Apprehension (not withstanding the many sorrowful Reflections he may be supposed to make upon it) his Case is very Desirable when set in Compare with that fool mentioned by our Saviour Luke 12th who when his Soul comes to be required of him shall be found only to have been laying up Treasure to himself, and is not Rich towards God for the former hath time and opportunity to provide himself of a far better and greater treasure that what he hath lost, whilst the other is past all hope in that Respect. The former hath Nothing to fear from the Rage of his Great Master, that the killing of his Body.

      But oh! What hath not the latter to fear from the Anger of an Incensed God, who had so often offered himself unto him as his Portion in order to his Everlasting Happiness, but was neglected & Slighted. And for what was the Gracious Offer Despised? Even for the Gratification of Sinful Lusts, or for the enjoyment of Lying Vanities which he very well knew he must soon leave and how soon who can tell? The Lord grant these Considerations may make suitable impressions upon all our Minds, to his Care I heartily commend you and with Dear Love to you, Duty & Dear Respects to all as due.

I am, your affectionate husband.
Tho Newcomen.

In 1920, a society was formed as a result of the James Watt centenary celebrations held in Birmingham the previous year. The object of the society, based at the Science Museum in London and still continuing to this day, is to support and encourage study and research into the history of science and technology and the preservation of records, both technical and biographical. Its subjects cover the whole field of industrial activities and the society maintains close links with other organisations with like interests. As a fitting tribute to the great inventor from Dartmouth, this society is called the Newcomen Society. A free translation of its Latin motto could be: 'That the future may learn from the past.'

In 1921, as a memorial to Thomas Newcomen, the Borough of Dartmouth erected a granite obelisk with an engraved bronze plate, which is now in the Royal Avenue Gardens, Dartmouth.

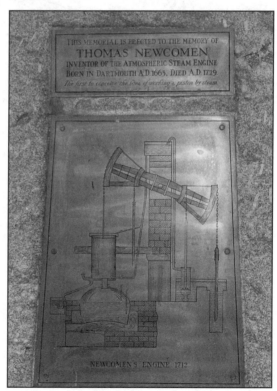

Fig. 82: The memorial to Thomas Newcomen erected in the Royal Avenue Gardens, Dartmouth.

The Newcomen Memorial Engine

In 1963, to mark the tercentenary of the great inventor's birth, the Newcomen Society had the opportunity to create a fitting memorial. A small atmospheric engine which had lain unused for almost fifty years became available and was erected in Dartmouth. This engine had been installed by the Coventry Canal Company to lift water from a well into the canal at Hawkesbury Junction, Warwickshire, in 1821. The engine was second-hand and was modified when installed on this canal pumping operation. The cast iron cylinder is 22" in diameter. Plates and flanges bolted to the cylinder indicate that the engine was operated in a completely different way when it was new. The pickle pot condenser, and also the valve gear, are later improvements. There is also an indication that the boiler to this engine was once positioned directly beneath the vertical iron cylinder. If these modifications are a true indication of how this engine originally worked, its date could possibly be as early as the 1730s.

For demonstration purposes, this engine is now hydraulically powered and these days it can be seen pumping water just as it used to do after being installed beside the Coventry Canal in 1821.

Fig. 83: Thomas Newcomen's Memorial Engine, Dartmouth, Devon. Illustrated by Nance Fyson, 1975.

CHAPTER 13

JAMES WATT'S IMPROVEMENTS TO THE STEAM ENGINE

1729 to 1775

Between these years, the atmospheric pumping engine was to become the standard means of removing water from both metalliferous and coal mines and its use was widespread throughout Great Britain and abroad. Engines had been steadily improved as their operating principles were better understood, while improved mechanical design allowed engines to be used with larger powering cylinders. One of the most noteworthy engineers who steadily developed the atmospheric engine during this forty-year period was John Smeaton (1724–1792).

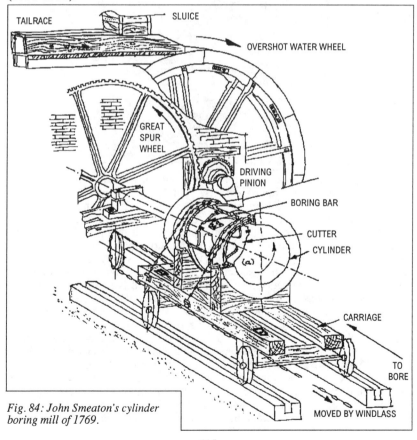

Fig. 84: John Smeaton's cylinder boring mill of 1769.

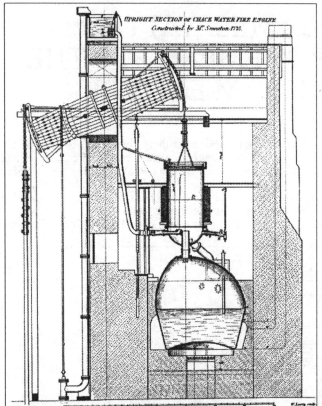

Fig. 85: The Chace Water Fire Engine, a large atmospheric pumping engine constructed by John Smeaton in 1775, used in a mine near Truro, Cornwall. Note the old spelling of 'Chace Water', instead of the modern version, 'Chacewater'.

Smeaton, who is better known for his design of the Eddystone lighthouse in 1759, made many improvements, mainly in the design and better proportioning of each mechanical part. The boring mill shown in Fig. 84 was used to machine the internal surface of the cylinder of the atmospheric engine. There was no great innovation in this mill – it was just a purpose-built machine, based upon the wooden pipe boring principle. The mill was driven through a large reduction gear by an overshot waterwheel. The cutting head on this boring machine had six cutting tools made from cementation steel (high carbon steel) attached to a strong axle, all driven by the great spur wheel. The mill was used by John Smeaton in 1769 and enabled him to build large engines such as that shown in Fig. 85. This engine was constructed by Smeaton in 1775 and is known as the Chace Water Fire Engine. The cast iron vertical cylinder had an internal diameter of seventy-two inches and it had a working stroke of ten feet. The details of the Chace Water engine are shown in Fig. 86 opposite and include the actual constructional dimensions.

As can be seen from these drawings, the engines were well proportioned and, when compared to Thomas Newcomen's engine of 1712, must have been capable of removing large volumes of water from the mine workings near Truro, Cornwall. The cylinder had a cross-sectional area of almost twelve

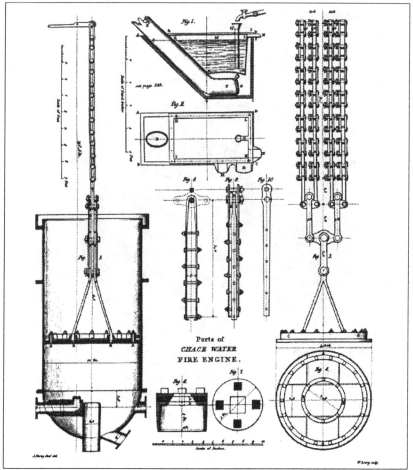

Fig. 86: Constructional details of Smeaton's Chace Water Fire Engine.

times the area of Newcomen's engine which had heralded the start of the industrial revolution almost 63 years previously.

Before proceeding to describe the engines designed after 1775, it is intended to give a short biography of the lives of James Watt and Matthew Boulton, the men who took the principle of the steam engine and developed it into a practical machine with more power to drain the mines of flooding water. Later, James Watt would design an engine which provided the rotary motion needed to drive the machines of industry which had previously had to rely on water and horse power.

James Watt (1736–1819)

James Watt was born in Greenock, Scotland on 19th January, 1736. His father was a general craftsman and worked as a shipwright, a ship's chandler and a builder. James Watt was a delicate child and was taught at home by his

Fig. 87: James Watt (1736-1819) from a portrait in the Scottish National Gallery.

mother before he went to school. After his primary education, he attended the Grammar School in Wee-Kirk Street, Greenock. He was not good at his lessons, but began to excel when he studied mathematics. While he was at home, he helped his father in the workshop and he had a workbench and forge where he made his own models. After leaving school, James joined his father for a short time but his father soon decided that he should go to Glasgow and prepare to become a mathematical instrument maker. Unfortunately, however, the only training available was with an optician who turned out to be a simple mechanic.

When James was working in Glasgow, he became friendly with the tutors from the Natural Philosophy department of the University. These men of intellectual distinction advised Watt to travel to London so that he could receive better training as a mathematical instrument maker. Later, these men were to have a profound influence on Watt's life.

The idea of going to London proved very difficult for Watt, as he had been told that wherever he went he would have to serve a bound apprenticeship of seven years. Not wanting to commit himself for such a long time, he sought the advice of a relative and was eventually offered a situation in the shop of a mathematical instrument maker in Cornhill. James Watt had to pay twenty guineas for this privilege and in return he received tuition for twelve months in his chosen trade. Consequently, Watt became rather short of money and had to rely on his father for an allowance.

The twelve months passed very slowly for Watt, but he was determined to gain as much knowledge as possible before he returned to Greenock. When eventually he returned to Glasgow in October 1756, his health was poor and he needed time to recuperate. He was anxious to make contact again with his former friends from Glasgow University and when he returned to the Natural Philosophy College of the university, the college was awaiting a consignment of fine astronomical instruments from Jamaica which, when they arrived, were found to be damaged and Watt was asked if he could restore them to their original working order. So that he could do this, he was given a room in the college and for the next six weeks he settled down to the job which he enjoyed.

The intellectual atmosphere of the university suited Watt's temperament and he opened a workshop in the university to make and sell mathematical instruments. This turned out to be a very shrewd move for him, as Watt had not served the established apprenticeship of seven years and would have found it very difficult to open a shop on private premises. But the tutors, appreciating Watt's talent, actually bestowed on him the title of Mathematical Instrument Maker to the university.

In his college workshop, Watt not only offered mathematical instruments for sale but repaired musical instruments such as flutes, guitars, harps and barrel organs. He also made a wind organ for his friend, Dr Black.

In 1759 James Watt became a partner with John Craig and they opened a shop in Saltmarket Street, Glasgow. This shop also had the facilities for making and repairing mathematical instruments, but Watt still retained his college workshop and continued to work for the university. From this beginning, James Watt was to become a most famous British engineer.

When you ask the questions 'Who invented the stationary steam engine?', and 'Who invented the locomotive?', the usual answers are James Watt and George Stephenson. Surprisingly, neither of these answers are correct. James Watt took the principles of Thomas Newcomen and refined them into a practical machine which could draw water from the coal and mineral mines. This refinement eventually produced the rotary motion so needed to drive the factory machinery where water was not plentiful.

James Watt can justifiably be said to be the greatest *improver* of the steam engine who ever lived, but he relied upon principles developed almost sixty years before.

George Stephenson developed his famous locomotive 'Rocket' from principles first established by Richard Trevithick in 1801.

Matthew Boulton (1728–1809) and the Soho Manufactory

Matthew Boulton was born at Snow Hill, Birmingham, on 3rd September 1728. His father was a small article manufacturer and the business was run from a small factory adjoining their house. It is thought that his father specialised in the manufacture of buttons and buckles and these small article manufacturers, who made products from metal, were called toy-makers.

At an early age, Matthew Boulton showed a keen interest in his father's

Fig. 88: Matthew Boulton (1728-1809).

business. On the death of his father, the whole business was transferred to him and under his management, the business grew and prospered.

He was an ambitious man with a vision and he designed a scheme to manufacture a large variety of hardware in just one factory building. His present site became too small and, so that he could carry out his plans for this expansion, he searched for a larger site. In 1761 he found such a site at Soho, only two miles from his Snow Hill factory. For this new venture, it was essential to have a good water supply, and at the Soho site there was already a water-powered mill.

After securing a lease for the site, Matthew Boulton became a partner with John Fothergill and each man put £5,000 into the venture. The new factory was soon begun, but the water mill proved unsatisfactory and was demolished.

When completed, the factory was a three-storey building and was the first of its kind for the manufacture of hardware. It also had an ample water supply to drive the new machinery. Fig. 89 shows an artist's impression of the Soho Manufactory, drawn in 1798. The old Snow Hill premises were closed down and the manufacture of the hardware was transferred to the new factory. With the development of this factory system, the business flourished and, within the first five years, two extensions were added.

The new factory came to be known as the 'Soho Manufactory' and was run

Fig. 89: An artist's impression, drawn c.1798, of the Soho Manufactory built by Matthew Boulton and John Fothergill.

by Matthew Boulton in partnership with several men, notably Fothergill and Scale. Some of the items produced at the Soho Manufactory were toys, silver and plate ware, pictures and document copying, buttons and buckles, coins and medals. In 1797, Boulton produced copper coins for the British realm and they became known as the Boulton coinage. However, the coinage was not accepted by the public because it was too heavy for common usage – for example, a two pence piece was made which actually weighed two ounces!

A most outstanding partnership was formed in 1775 when James Watt joined forces with Matthew Boulton to produce Watt's improved steam engines. The water power at the Soho Manufactory became inadequate and a new source of power had to be found to drive the machinery. Matthew Boulton thought that James Watt's engines would be suitable to provide this extra power as they were single action pumping engines and they could be used to pump water into the tail-race of the waterwheels, which would provide the extra rotary power. Boulton and Watt's engines were used for the draining of mines in many parts of the country and the Cornish tin mines were noted for their use of these single action pumping engines.

James Watt only designed these steam engines. The complete engines were never made at the Soho Manufactory but the component parts were made by other manufacturers, such as John Wilkinson, who made the cylinder and the main castings at his Bersham Ironworks.

William Murdock (1754–1839)

The one person who is inextricably linked with the success of the Boulton and Watt partnership is William Murdock. He was born at Bellow Mill, Old Cumnock, Ayrshire, on the 21 August 1754. His father was a millwright, and from an early age the young Murdock showed a good aptitude for mechanics, but the opportunities for a young man of such abilities were few in his home

Fig. 90: William Murdoch (1754-1839) from an oil painting in the Birmingham Art Gallery.

town. He therefore journeyed south in 1777 in the search for suitable employment.

By this time, the news of developments at the Soho Manufactory had reached Scotland, so Murdoch went to Birmingham in the hope of obtaining employment at the new factory of Boulton and Watt.

Fortunately, on his arrival Murdoch met only Matthew Boulton because, although born in Scotland, James Watt did not consider that country a good training ground for engineers. Matthew Boulton was a shrewd judge of character and he immediately took a liking to this young man.

The classic story of their interview has often been recounted. As Boulton questioned him, the young man stood before him nervously running the brim of his hard hat through his fingers. Boulton became fascinated by the hat until he could not keep his eyes off it. It was like no other hat that he had ever seen and, instead of a cloth covering, it looked as if it had been painted.

"That seems to be a curious sort of hat," he said at last. "What is it made of?"

"Timmer, sir," replied Murdock promptly. ('Timmer' is a colloquial word for 'timber' – *editor.*)

"Timmer!" exclaimed Boulton, "But how was it made?"

"I turned it myself, sir, in a bit of lathey of my own making," was the reply.

Boulton realised that such an unorthodox job would require a lathe which could turn oval shapes and that a man who could make and use such a machine was no mean mechanic! Murdock was promptly engaged and soon Boulton's judgement was proved correct.

Murdock was first employed as a patternmaker, but this work was followed by erecting and modifying pumping engines in the Midlands. His first job on a large scale was to erect a pumping engine at Wanlockhead, Dumfries. This was an engine with a 36" cylinder, designed to remove the flooding water from a coal mine.

With this work complete, Matthew Boulton had more ambitious plans for the young man from Scotland. In September 1779, he was despatched to Cornwall where he was to make his home for the next 15 years.

The first Boulton and Watt engine had been erected in Cornwall at Wheal Busy in 1777. So successful was the Watt engine that by 1783 James Watt actually claimed that only one Newcomen engine was left in Cornwall. This was an exaggerated claim, but it showed how superior his engine, with a separate condenser, was compared to the engines which had been used until then.

William Murdock worked tirelessly on perfecting these engines during his stay on Cornwall. He made many modifications, and on occasions he changed the specification of an engine without any consultation with Matthew Boulton or James Watt. This led to many disagreements among the three men – the two original partners were always afraid that Murdock would forsake them and set up business on his own.

In 1784 he actually demonstrated a steam powered carriage. The story is told that he tried this carriage in a churchyard near to his home in Redruth. However, Matthew Boulton persuaded William Murdock not to take out a patent for this carriage and, like many of William Murdock's ideas, James Watt actually patented the steam carriage for himself. The original carriage is now in the Birmingham Art Gallery.

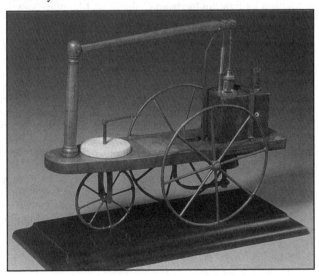

Fig. 91: Murdoch's carriage made in 1784, now in the Birmingham Art Gallery.

123

Fig. 92: James Watt studying the Newcomen model.

Of his many inventions, the patent taken out by Murdoch in 1791 is the most noteworthy, because this patent dealt with the burning of coal-gas as an illuminator. He had perfected a method of making gas in an iron retort in his back yard. In 1792 he successfully lit up his house and offices in Cross Street, Redruth – the first buildings in the world to be illuminated in this way. This lighting became known as 'illumination without wicks'.

In 1797, William Murdock returned to Birmingham and carried out further experiments on the production of gas and in 1802, he staged a public exhibition by lighting the front of the Soho Manufactory with two gas lights.

The Separate Water Cooled Condenser

The development of the separate condenser was the greatest single *improvement* ever made to the steam engine! In 1763, when James Watt was working as an instrument repairer at the Natural Philosophy College in Glasgow, he was asked to repair a model of Thomas Newcomen's pumping engine. This repair was eventually completed and everyone was satisfied, but Watt was not happy with the amount of fuel the model consumed. He thought that too much energy was needed for the amount of actual work done. The condensation of the steam to create the powering vacuum was taking place within the working cylinder and, operating in this way, the cylinder was heated and then cooled for each cycle of the engine.

It became very obvious to James Watt that a great improvement could be made if this operating vacuum could be formed *away* from the main cylinder. Ideally for a good vacuum to be formed, condensation should always take place in a cold vessel. By 1765, he was completely familiar with the working principles of the atmospheric engine and, with this in mind, he set himself the task of trying to improve its working efficiency.

When he was taking a fine Sunday afternoon walk in May 1765 and thinking of nothing but improving engine efficiency, the idea came into his mind that steam was an elastic body – it would rush into a vacuum and, if the communication was made between the cylinder and the exhausted vessel, it would rush into it and might there be condensed without cooling the cylinder.

So, with such inspiration, the principle of the separate condenser was born. The principle of the separate condenser was soon settled by experimental models which Watt constructed to test his theory.

After the conception of the condenser, James Watt had taken full employment in surveying and civil engineering and, because of his family commitments, was frightened to start on such a speculative venture. Only in his spare time did he continue to experiment with his separate condensation theory and all that was necessary to make it work.

Nevertheless, in 1767 Dr John Roebuck (1718–1794), a man who had large interests in manufacturing industry, took up Watt and his invention. He supplied the financial support and put pressure on Watt to get practical results. However, progress was very slow because Watt was reluctant to lose his secure form of income.

By the year 1769, a patent had been granted to Watt for an engine to be built at Dr Roebuck's property at Kinneil, Scotland. This engine turned out to be disappointing, because the money ran out before the development was completed.

Dr Roebuck was in debt and Matthew Boulton paid off his debt of £1,200 and so acquired Dr Roebuck's interest in the Kinneil engine. The engine was then re-erected at Matthew Boulton's Soho Manufactory in Birmingham and, by 1774, James Watt had the engine working successfully.

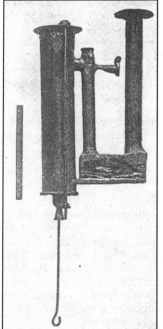

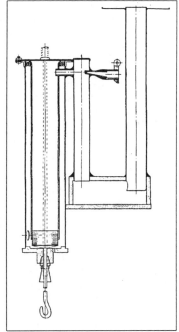

Fig. 93, left: James Watt's model of the separate condenser.

Fig. 94, right: A sectional drawing showing how this condenser most probably worked in 1769.

An extension to the original condenser patent was taken out in 1775, almost ten years after its original conception and at the same time he joined forces with Matthew Boulton, thus forming one of the world's most successful and lasting engineering partnerships. He now had the opportunity of seeing his ideas put into practice without the financial constraints of the past.

The James Watt Patent

The title of the patent taken out by James Watt was 'New Invented Method of Lessening the Consumption of Steam and Fuel in Fire Engines' and was filed as patent number 913 on the fifth day of January 1769.

Extracted principles from this patent are as follows:

'First, the vessel in which the powers of steam are to be employed to work the engine, which is called the cylinder in common fire engines, and which I call the steam vessel, must during the whole time the engine is at work be kept as hot as the steam that enters it, first, by enclosing it in a case of wood or any other materials that transmit heat slowly; secondly, by surrounding it with steam or other heated bodies; and, thirdly, by suffering neither water or any other substance colder that the steam to enter or touch it during that time.

'Secondly, in engines that are to be worked wholly or partially by condensation of steam, the steam is to be condensed in vessels distinct from the steam vessels or cylinders, although occasionally communicating with them. These vessels I call condensers, and whilst the engines are working, these condensers ought at least to be kept as cold as the air in the neighbourhood of the engines by application of water or other cold bodies.

'Thirdly, whatever air or other elastic vapour is not condensed by the cold of the condenser, and may impede the working of the engine, is to be drawn out of the steam vessels or condensers, by means of pumps wrought by the engines themselves, or otherwise.

'Fourthly, I intend in many cases to employ the expansive force of steam to press on the pistons, or whatever may be used instead of them, in the same manner as the pressure of the atmosphere is now employed in common fire engines. In cases where cold water cannot be had in plenty, the engines may be wrought by this force of steam only, by discharging the steam into open air after it has done its office.'

'Lastly, instead of using water to render the piston or other parts of the Engines air and steam tight, I employ oils, wax, rosinous bodies, fat of animals, quicksilver and other metalls, in their Fluid State.

James Watt

Sealed and delivered
in the presence of
Coll. Wilkie, Geo. Jardine, John Roebuck'

Watt's idea of the separate condenser was the greatest single improvement to the steam engine of all time and he was cleverly protected by this patent, although the patent itself limited technological progress for almost 25 years. James Watt also included in this patent (item four) legal coverage of the expansive force of high pressure steam. On his engines, he only used the force of steam at a low pressure, but the coverage of this fact in the patent prevented other engineers from designing engines which would have developed more power and it was not until the year 1800 that engineers were allowed a free hand to design a new generation of steam engines.

CHAPTER 14

THE SMETHWICK ENGINE
OF 1779

Boulton and Watt

The atmospheric pumping engine had worked from Thomas Newcomen's original design of 1712 for almost seventy years. Engines had become more powerful, but this extra power was the result of better engineering principles – what had not changed was their powering force of a vacuum which was still created by condensing the steam within the main powering cylinder of the engine. However, this was all about to change now that James Watt had perfected his separate water cooled condenser. He was now able to exploit this great improvement, secure in the knowledge that a patent had been granted with all its legal protection.

Matthew Boulton and James Watt developed into one of the greatest engineering partnerships of all time. Their engines were eventually used throughout the world, because it had been found that an engine powered with a separate condenser doubled the power output for the same amount of coal used.

Boulton and Watt designed and made these improved engines, but they also charged the customer an annual premium, which was based upon calculating how much more coal would be needed if a standard atmospheric engine based upon the Newcomen principle had been used.

The first fully documented engine made by this partnership was designed for the Birmingham Canal Navigation Company, and is described in detail on the following pages.

The Smethwick Engine

By the mid-1770s, the use of an arm of the Galton canal was becoming severely restricted to canal traffic due to lack of water. The flight of locks at Spon Lane which dropped the canal to the Birmingham level consumed more water from the top canal level than the existing reservoirs could supply, so a solution was required if canal traffic was to increase.

James Watt's newly perfected separate condenser enabled much larger and more powerful steam engines to be made, but could enough water be returned from the bottom of the flight of locks to replenish the canal with sufficient water to allow the traffic to be increased?

It was in August 1776 that the canal company decided to seek Matthew Boulton's expertise on this problem. His advice was to erect a trial engine near Spon Lane and pump the water to a higher level in the lock flight. Samuel Bull, the canal company's engineer, went to visit James Watt in February 1777 and an engine was ordered. By early April 1778, the engine was complete and working. However, the Spon Lane engine, as it became known, was only small, delivering about four horsepower, but the important point had been established that it was possible to pump water to a higher canal level by a fire or steam engine. This engine was the first engine in the world to be used by any canal

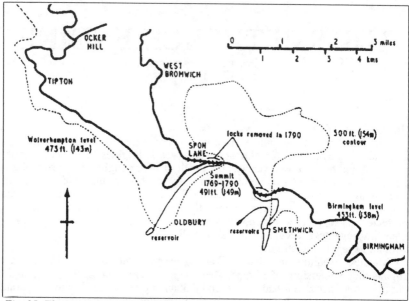

Fig. 95: The canal system west of Birmingham in the eighteenth century, also showing the 500 ft (154m) contour.

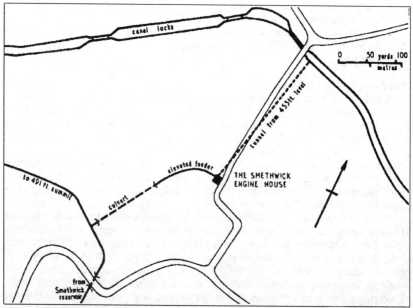

Fig. 96: Where the Smethwick Engine originally stood in 1779.

company for water conservation and became known as Navigation Engine Number One.

Within two months of the commissioning of the Spon Lane engine, the members of the canal committee were producing more ambitious plans. They asked Samuel Bull to look for the best site to erect an engine to lift water from the bottom to the top canal level, a height of 38 ft. Lifted to this height, the water could flow naturally back to the top lock to replenish the canal with much needed water. If this ambitious plan could be carried out, the water level would remain relatively constant and the canal company could indeed increase the canal traffic. Samuel Bull was also asked to recommend the best type of engine to use for this demanding job.

The site selected for the engine was on the intersection between Rolfe Street and Bridge Street, Smethwick. The two maps, Figs. 95 and 96, show

Fig. 97: A general arrangement drawing showing how the Smethwick Engine lifted the canal water to flow back to the top lock.

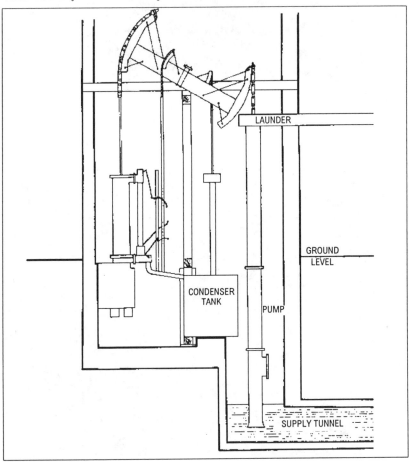

Fig. 98: The Smethwick Engine foundations at the top of Bridge Street, Smethwick. This photograph was taken in 1990.

Fig. 99: Another view of the foundations.

Fig. 100: The bridges show, by being built at different levels, how the canal water level was lowered in 1790.

where the engine originally stood, though all that now remain are the original foundations, conserved for posterity by a modern fabricated building. The photographs of the site seen at left were taken in 1990 and clearly show where this large pumping engine stood upon its original brickwork foundations.

Once more a Boulton and Watt engine was chosen to lift the water, but this time the engine was a large one with many unique operating features. The vertical powering cylinder was 32" in diameter, with a working stroke of 8 ft. The feeder tunnels to the engine site were started on the 1st August 1778 and the installation was completed by June 1779, when James Watt was able to carry out proving trials on the engine, which made twelve pumping strokes per minute and, on each stroke, lifted almost 160 gallons of water which flowed naturally to refill the canal at the summit. Following the Spon Lane engine, the new engine became known by James Watt as Navigation Engine Number Two and it continued to pump water until early 1790.

In 1789, a new feeder was constructed from the engine to the 473 ft level, above the third lock, in preparation for lowering the level of the canal by 20 ft. Therefore, in 1790 the original water lift was reduced and the result can now be seen in the unusual photograph, taken in 1990, of the different bridge levels over the canal summit which graphically show the lowered canal level.

Many modifications were needed to the original 1779 installation. For example, the diameter of the pump barrels was increased from the original 24" to 29" diameter and the discharge launder to the summit was lowered to just above the floor level of the engine house. The engine continued to work after this major reconstruction until 1891, which is thought to be the longest period of time a steam engine has worked anywhere in the world!

Fig. 101: This historic photograph shows the Smethwick Engine house being demolished in 1897.

The only surviving photograph of the Smethwick Engine at the top of Bridge Street (Fig. 101) was taken in 1897 where the two workmen can be seen demolishing the engine house. Even in the last century, the Smethwick Engine was thought to be a very important historical artefact and so it was re-erected at the canal company's depot at Ocker Hill, Tipton. The engine was not used at this location, but just preserved for ceremonial purposes.

In the late 1950s the Ocker Hill depot closed and in 1959 the Smethwick engine was presented to the Birmingham Museum of Science and Industry, but it was not until 1983 that the engine was reassembled and could be seen pumping water – the first time that the engine had been steamed since 1891, though this time the steam was generated by the museum's boiler.

With the closure in 1997 of Birmingham's Museum of Science and Industry in Newhall Street, the Smethwick engine is now held in storage until a suitable display can be provided within the museum's new site at Digbeth.

CHAPTER 15

THE SMETHWICK ENGINE IN MINIATURE

The story now continues of how the model of the Smethwick Engine was researched and made. With the model's construction will be described the many unique features developed by James Watt, because this was the beginning of a new generation of engines operating on principles never before used with steam power.

An engine which was built in 1779 and worked continually until 1891 (usually 24 hours per day!) must have had many replacement parts incorporating the latest developments in technology. Therefore, in 1891 it is probable that the engine would not have appeared as it was originally designed in 1779 by James Watt.

After months of study and consultation with both the Newcomen Society and the Birmingham Museum of Science and Industry, it appeared that the Smethwick Engine was the next logical step in my quest to recreate the most technologically advanced engines which were built in the eighteenth century, and the intention was to construct the model as it would have been seen more than two hundred years ago.

In November 1989 an appointment was made to have a talk with Jim Andrew at the Birmingham Museum of Science and Industry to establish what documented information the museum held on the Smethwick Engine, and if anyone would be able to help to establish how this engine would have appeared on completion in 1779. At this meeting a surprising discovery was made – Mr Andrew had personally researched this engine over the past eleven years and he agreed to share this research to enable an accurate model of the engine to be made.

The model took almost four years to complete, and during this time 4,100 hours were spent on construction. Mr Andrew was consulted on the smallest of details and he also supplied many scaled sketches to enable the model of the engine to operate without the modifications which had taken place during its long working life.

The planning really started after taking many measurements of the engine in the Birmingham museum. These revealed how big the finished model would be at my chosen scale of $1/16$ full size. The favoured method when planning such a large project is to take numerous photographs and, on each exposure, to place a rule within the field of view. Using this method, it is very easy to scale the photographs and thus minimise the number of detail drawings.

An outline arrangement drawing was completed of the engine house, showing how each part fitted together. The only surviving original drawing which was able to help is shown as Fig. 102 on the next page. This has been redrawn for clarity, because the original drawing is too faint for reproduction.

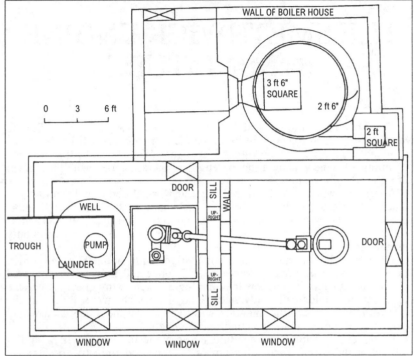

Fig. 102: This drawing shows how the Smethwick Engine was positioned within its original building in 1790. This drawing is taken from the only surviving drawing of 1779 known to exist.

A Start is Made

After much deliberation, it was decided to make all the mechanical and wooden components which were known *not* to have changed since 1779. It is generally accepted that the engine in the Birmingham Museum still uses the original beam and all its mechanical attachments. This large baulk of English oak measures almost 24" square and has a length of 22 ft. In 1779 this beam was supplied by Glace & Higgs for £11-1s-9d.

So as to give the correctly scaled appearance to this beam in miniature, the chosen wood was Japanese oak which is of a similar colour and hardness to English oak but has a much closer grain. The photograph, Fig. 103, shows the finished beam within the roof space of the miniature engine house.

Positioned in this way within the engine house, the beam appears unusual. The accepted practice at this date was to make one gable wall thicker and have the pivot support bearings within the actual wall. However, James Watt was questioning all the established principles at this time. This positioning of the beam pivot within the building does not appear to have been repeated and after this engine, Watt reverted to the much simpler method with the well head on the outside of the buildings.

With the model of the main operating beam complete, the building sizes were starting to emerge. My method of making replica engines is to make all

Fig. 103: The miniature Great Beam within the model engine house. In 1779, the original beam measured over 22ft long and was made from English oak.

repetitive parts first and finish with the more pleasing and interesting items. In this way, the long hours spent on construction do not appear too daunting!

With this philosophy in mind, a start was made on the chains which were used at each end of the arch heads, which James Watt sometimes called 'horses heads'. The photograph below shows the detail involved – every link is hand forged and the split cotters are located in rectangular slots 2mm by 0.5mm wide. When asked how were they made, the only answer is carefully, slowly and very patiently!

Fig. 104: The arch head and chains of the model engine. Each chain link and each square nut and bolt has been made by hand.

CHAPTER 16

THE POWERING CYLINDER OF THE SMETHWICK ENGINE

In 1803, Boulton and Watt overhauled the Smethwick Engine and a major modification took place in which the original 32" diameter cylinder was replaced with one of 33" diameter. To operate this new cylinder, all the valve operating mechanisms were also changed. Therefore, when Boulton and Watt had completed the overhaul, the engine no longer bore any resemblance to how it had originally appeared in the eighteenth century.

Fig. 105: Air and hot water pumps from the separate water cooled condenser.

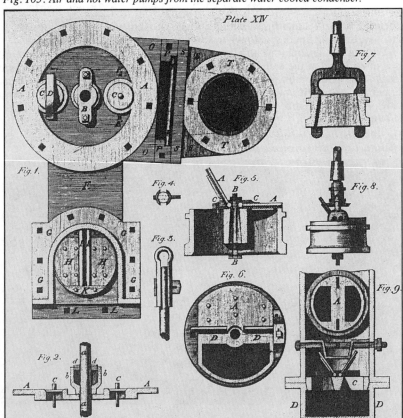

Fig. 106: The main cylinder and cylinder head showing the first stuffing box ever fitted to a steam engine.

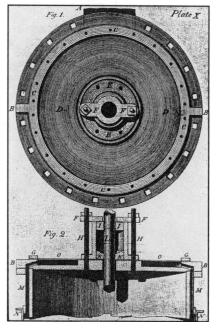

As no detail drawings survive of the original cylinder, the only path now open to me was to study all the written research of the period to see if it was possible to rediscover how James Watt had originally designed the cylinder in 1779. The most important written evidence was a compilation of the costs incurred when the engine was originally made.

Jim Andrew found the original lists of the costs at the Public Record Office at Kew. These records were to provide invaluable information and settle any uncertainty during the construction of the miniature engine. The item on this list which gave the initial lead was an entry for twelve cylinder plates costing £18-8s-7½d. These segmental plates were made by Dearman Winwood & Freeth – this company made the wooden pattern and cast these plates in iron.

The cylinder plates would have been fitted to create a cavity around the main powering cylinder and then filling this void with the incoming steam. In this way, the cylinder would remain at a constant high temperature. This was James Watt's theory for the ideal running of his engines with the separate water cooled condenser. Condensation of the steam was taking place in a cold vessel, so the cylinder was maintained to a high temperature – the exact wording of his original 1769 patent!

When closely studying the parts lists and prices from the Records Office, an important discovery was made. Figures 105 to 108, from H W Dickinson & R Jenkins' book, *James Watt and the Steam Engine,* are drawings of the Smethwick Engine cylinder! The book does not suggest this, but there cannot be any other conclusion because the number of components are the same and James Watt only made one design of engine in that year (1779).

The Smethwick Engine was the first engine in the world to use both the expansive force of steam and a vacuum created in a separate water-cooled con-

Fig. 107: The piston, showing how a seal of hemp, lubricated with tallow and olive oil, was made with the cylinder wall.

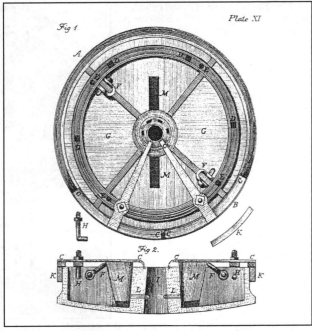

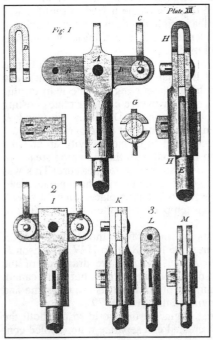

Fig. 108: A detail drawing showing how the connections were made between the piston rod and pump rods on to the arch head chains.

Fig. 109: The powering cylinder of the model engine. The Smethwick Engine had a steam jacket made from cast iron segments bolted together and it is thought that only a small number of engines were ever constructed in this way.

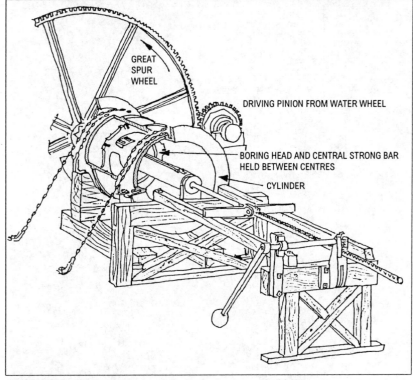

Fig. 110: John Wilkinson's patented cylinder boring mill of 1774.

denser to power the piston. Steam at about 3 psi was directed to the top of the piston, then a large volume of steam were fed into the condenser where, after condensation, a vacuum was created which forced the piston to the other end of the cylinder bore. The weight of the beam and the heavy pump rod rods then returned the piston back to the top of the cylinder, thus completing one cycle of the engine.

To achieve this revolutionary operating cycle, the engine was the first to use a system which became synonymous with James Watt – the three valve operating system.

Valve One allows steam at 3 psi to push down on the piston top (*expansive working*),

Valve Two allows the steam into the condenser where, after condensation, a vacuum is created, and

Valve Three directs the created vacuum to the underside of the piston (*atmospheric working*).

This method of moving the piston with the expansive force of steam and a vacuum brought about other problems. The neat solution to one problem is to be seen in Figure 106. The piston rod can be seen passing through a gland in the cylinder head, making this the world's first stuffing box on a steam engine cylinder, as no cylinder before the Smethwick engine would have needed a cylinder head – as they were only ever worked by a vacuum on the underside of the piston.

As can be seen from the photograph, Figure 109, the powering cylinder of the Smethwick Engine was complicated by the twelve segmental plates bolted together to form a steam jacket. Even the cylinder head had a cavity for steam to keep it hot, this cavity being sealed by a copper plate clamped in position by a large annular ring fixed to the top of the cylinder.

The cylinder and all the pipework was cast by John Wilkinson at his Bersham Ironworks. Wilkinson's boring mill, shown as Fig. 110, was a development of John Smeaton's boring mill and must have been used to machine the bore of the cylinder used on the Smethwick Engine in 1779. Cylinders needed their internal surfaces finished more accurately than would have been possible on Smeaton's boring mill, as the piston had to be sealed by close tolerances, rather than with water flooding over the top of the piston as in a completely atmospheric engine.

Cutting Tools on Water Powered Mills

The cutting tools were made from high carbon steel, suitably hardened and tempered. This steel was produced by either the cementation process or the crucible process. The cementation process of making high carbon steel was a well-established method which had been in use for many centuries, while the crucible process was a relatively modern introduction invented by Benjamin Huntsman in 1740, though this latter process did not come into general use until the 1770s.

In the early years, the boring mills used cementation steel imported from Sweden and Russia for the cutting tools. These two countries had developed the cementation process to perfection. Cementation is the conversion of wrought iron by means of carbonisation into steel, and this was brought about by subjecting wrought iron to carbonaceous materials at a very high temperature over an extended period.

Wrought iron bars were heated by a combination of charcoal and wood, and maintained at a temperature of 900°C for up to a week. The high temperature soak was dependant on the type of steel required, ie. the longer the soak, the higher the percentage of carbon in the steel produced – for cutting tools, 1.2 percent or slightly more was required. However, the production of the steel was very expensive because large quantities of fuel were required to maintain the high temperature over such a long period. The conversion of 1,000 lb of wrought iron into steel took one hundred tons of charcoal and wood, so a great number of trees had to be felled.

In an attempt to reduce the high cost of production, coal was used for the first time, with great success, with the coal being exported from England to Russia to heat the cementation furnaces and the coal method of conversion soon became the only way of producing cementation steel in Britain during the eighteenth century.

Unfortunately, steel produced by the cementation process had a tendency to swell, resulting in a blistery appearance, and so was called 'blister steel'. The bubbly appearance was due to the reaction between the carbon and the residual oxygen at the time of carbonisation. After cementation, consolidation had to take place and this was achieved by forging, before being formed into cutting tools, followed by hardening and tempering.

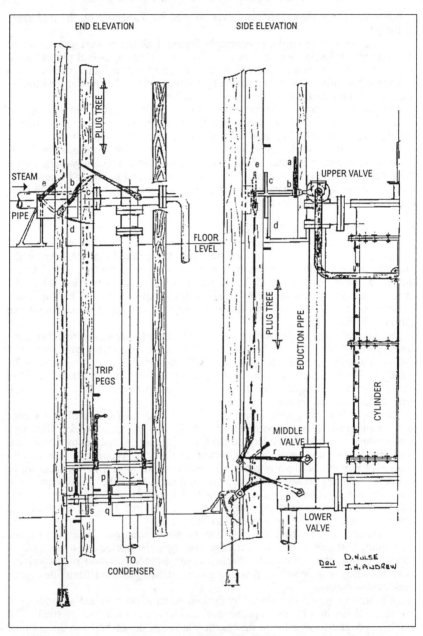

Fig. 111: The valve operating leverage based upon the engine built by Watt for Peter Colevile.

CHAPTER 17

CONSTRUCTIONAL DETAILS TO THE SMETHWICK ENGINE

The Valve Operating Leverage

Figure 111 shows the arrangement of the levers controlling the flow of pressurised steam and also the application of the vacuum to the underside of the piston of the Smethwick Engine. The drawings are taken from an engine designed for Peter Colevile of Torryburn, Fifeshire, by James Watt in 1776 – this engine has drawings reproduced in *James Watt and the Steam Engine* by Dickinson and Jenkins, 1927.

Although the Scottish engine was only powered by atmospheric pressure to the top of the piston, the drawings give a good indication of how the Smethwick Engine would have been controlled, as this valve operating arrangement fits neatly into the Smethwick Engine foundation details, giving confidence that the lever arrangement fitted to the model engine was not too far removed from how James Watt intended it to work in 1779.

The Powering Cylinder in Miniature

From the initial planning, it was intended that when the model engine was complete it should be able to be seen working just as James Watt had intended it in 1779. However, it was necessary to power the cylinder in *both* directions – thus making the powering cylinder of the model double-acting – as gravity alone would not have overcome the friction of the piston on the miniature engine and so return the piston automatically to the top of the cylinder for the next pumping stroke.

Lubricated compressed air powers the piston in both directions. Because the piston seals are very difficult to replace, an admission here seems appropriate. The seals on the model engine are made from a very modern material – PTFE – and the bore of the cylinder was finally finished to a diameter of 50mm by honing.

There are no castings used on the model engine – every part was machined from solid mild steel, while the valve chests are attached to the cylinder by soft soldering.

The 9" long, 3½" diameter mild steel billet used for the cylinder weighed 24 lb before any metal was removed, but it had been reduced to 6 lb when boring was finished! The external surfaces of the cylinder were then 'roughed up' by being held against a high speed linishing belt, removing all the tool marks from turning and giving an appearance of a fine sand casting. The nozzle housings were then soft soldered in position and the finishing operation was to spray the cylinder with a grey metal primer, followed by two coats of matt black paint. This is not an authentic finish, but does prevent any future discolouration from corrosion, and matt black paint is the generally accepted finish for conservation of exhibits displayed in museums, where corrosion could become a problem.

The Pump Barrels

The original pump barrels fitted to the Smethwick Engine in 1779 had an internal diameter of 24", and these were bolted together to form a continuous cylinder 38 ft long. The barrels on the miniature replica of the engine have an overall length of 28" and an internal diameter of 1½". Five individual pieces of mild steel were used to make the model pump cylinder, each piece held to the next by eight square headed nuts and pins, which pass through flanges at the end of each barrel.

With many of the mechanical and wooden parts to the miniature engine now made, a start could then begin on the planning of the complete engine house model, as the building sizes were beginning to emerge through close study of the remaining evidence of the Smethwick Engine and also from the preserved foundations at Smethwick.

Fig. 112: The Smethwick Engine house in model form, showing where the water flowed into the vertical brick lined well.

144

The Remaining Evidence

The photograph showing the Smethwick engine house being demolished in 1897 is the only record which shows the engine positioned within the engine house, so this photograph had to be studied closely to establish many of the building sizes and construction details.

At this stage it was thought that the only way to progress was to visit and measure the foundations at the top of Bridge Street, Smethwick. Numerous photographs were also taken. By measuring and recording the size of the bricks used in the foundations of the building, it was easy to calculate the size of the engine house without taking all the dimensions. The sizes could be established by counting the rows of bricks, and multiplying by the known brick size.

From the surviving foundations, it was found that the surrounding walls were very thick, so a preliminary calculation was then made as to how many miniature bricks would be needed for the project. Then the clay powder for use in the miniature brickmaking machine was purchased and the cycle counter on the brick machine set to zero.

A decision was made that all the bricks for the project should be prepared and fired before any construction of the model engine house was started, so the following three weeks were taken up on the laborious task of compressing the clay powder into the small clay bricks. The brick machine worked very well and completed the task after running for approximately fifty hours, producing 45,000 bricks (on completion, only 38,000 bricks had been used!).

An Interesting Constructional Detail

A close study of the foundations to the Smethwick Engine revealed a most unusual constructional detail. Most powering cylinders were fastened into their working position with bolts 'grouted' into holes in the foundation stone with lead, but the Smethwick Engine was held in its working position by four 4 ft long wrought iron bolts of almost 2" diameter, made to pass through the large foundation stone beneath the cylinder. At the lower end of the bolts, there was a wrought iron plate approximately 1½" thick – the bolts passed through this plate and were then prevented from lifting by four large tapered keys. Nuts at the top of the bolts were then tightened. This unusual method of holding down the cylinder meant that should any of the fastening bolts fail, a replacement could easily be fitted.

The cylinder of the miniature engine is also held in position by long bolts with tapered keys at their bottom and these bolts can be seen readily in the sectional view in this area of the model engine house (Fig. 112).

The Building to House the Replica Engine

With many of the mechanical parts now completed, a start could be made on the planning and the eventual construction of the miniature engine house for the model of the Smethwick Engine. A drawing showed how the engine house would have looked in 1779, but a method was required to determine accurately the final dimensions of the building so that the engine would work smoothly when securely bolted into its working position. The dimensions to the building were calculated by assembling all the mechanical parts into their working position using numerous clamps and brackets, all supported from a welded steel frame securely bolted to the workshop floor. For reference, all the parts remained in this position whilst the building was constructed, with the final dimensions being transferred from the assemblage of parts and used to construct the miniature engine house.

The construction of the whole building was so complicated and involved that it is almost impossible to describe! A point of interest is that the foundations were built not from the lowest point of assembly to the working floor level as would be expected, but in fact built from the floor level *down*! This was achieved by inverting the whole structure to complete the foundations before turning it right side up again to construct the building above the floor level.

Wooden formers were made to hold the miniature bricks in their final position whilst the adhesive set. The flying arches were made by using a wooden former just as a bricklayer would have made the arches in 1779. The sectional representation through the underlying ground down to the water inlet was made by glueing and screwing together sheets of chipboard, with the final sculptured appearance being achieved by a high powered sander. This produced a surface which was ideal for a finishing layer of car body filler. The surface area was then hand painted, with numerous colours blended together to give a representational cross-section through the earth. This would show the working floor level down to just below where the water flowed into a vertical brick-lined well.

The Boiler to the Smethwick Engine

The supply of steam at 3 psi was generated in a haystack boiler with a diameter of 10 ft – in some parts of the country these boilers were known as beehive boilers. To fill the boiler to its optimum working water level, 2,400 gallons of water were needed. Fig. 113 below shows how these boilers were rivetted together from the many wrought iron plates. In 1778 all the materials

Fig. 113: An original haystack boiler once used at Basset Pit, Denby.

to make the original Smethwick engine's boiler were supplied by William & Stephen Hipkiss. This company charged £57-1s-2d for these materials and £14-16s-0d to rivet all the plates together, while the carriage charge to Rolfe Street when the boiler was completed was £2-2s-6d.

Many of the early Watt boilers were made of copper but the list of parts deposited in the Record Office at Kew clearly states that the Smethwick Engine had a boiler made from wrought iron plates.

In the early days of the Soho Manufactory, good wrought iron plates were difficult to produce. The method used was to forge wrought iron billets into the required flat sheets, but success varied depending on the type of wrought iron used. English billets were almost impossible to use because of their brittle nature, and a much greater success was assured by using billets imported from Sweden or Russia, both countries which produced a ductile and easily forged iron.

Because of this ductility, Russian slabs, as they came to be known, were widely used. Iron plates of $^1/_4$" and $^3/_8$" thickness were in common use at this time, and as the Smethwick boiler was of such a large size, the thicker plates were probably used. All of these plates were held together by $^5/_8$" diameter rivets.

The following paragraphs are the actual directions issued by Boulton & Watt for boiler making:

'In making the boiler you should use rivets between five-eighths and three-quarters of an inch in diameter. In the bottom and sides the heads should be large and placed next to the fire, in the boiler top the heads should be on the inside. The rivets should be at two inch centres and their centres should be one inch from the edge of the plate.

'The edges should be accurately cut, both inside and out. It is impossible to make a boiler top truly steam tight which is done otherwise.

'After rivetting, the edges of the plates should be thickened up, this is done by a blunt chisel about a quarter of an inch thick, impelled by a hammer of three or more pounds in weight around the whole of the joints. After caulking, the joints should be wetted with a solution of sal ammoniac in water, or rather urine, which by rusting, will help to make the joints steam tight.

'After the boiler is set in the brickwork, all the rivet heads are painted with whiting and linseed oil.'

From reading these instructions, one realises how skilled the craftsmen of this time were. They could produce a safe steam tight vessel, by using hand methods only. The production of the rivet holes must have been a skilled undertaking, as they were produced by using a hand held punch which was struck with a large hammer.

The Boiler in Miniature

The boiler shell of the Smethwick engine in miniature is made by segments of bright shim steel accurately riveted together. Each segment is 0.020" thick. To form the segments into the final convex shape, a concave wooden former was made to the exact external dimensions of the original boiler at the replica's scale of $^1/_{16}$ full size. Rivets slightly larger in scale were chosen to give added interest to the finished dome positioned within the lean-to building at the rear of the main engine house.

With the boiler shell now complete, the attachments such as the safety

Fig. 114: The model haystack boiler and boiler house of the Smethwick Engine.

valves and the water feed tank could now be fitted. These features were almost identical to the valves which were to be fitted to the Boulton and Watt Lap Engine of 1788.

With the boiler shell set into the miniature brickwork of the boiler house, the model Smethwick Engine and its boiler looked as if James Watt had just

Fig. 115: The rear of the model engine house showing some of the 38,000 reduction fired bricks used in its construction.

COMPANIES INVOLVED IN THE CONSTRUCTION OF THE SMETHWICK ENGINE AND ASSOCIATED WORKS

B. Hughes & Co.	Labour
B. & R. Blythe	35 yards of flannel
Blastus Hughes & Co.	Labour
Boulton & Watt	Licence for new engine
Boulton & Watt	Sundry materials
Dearman & Co.	Cast iron work
Dearman Winwood & Freeth	Cast iron
Edward Bates & Co.	Boring
Edward Turvey	Boring
Glace & Higgs	Beam
Green & Co.	Sinking foundatons
Green, Birch & Co.	Labour at the embankment
Grigg	Ironwork
H. Hipkiss	Carriage of bricks
Hannah Hipkiss	Hauling
Hipkiss	Boiler & carriage
Isaac Walton	Scaffold
Isaac Tonks	Nails
James Brough	Six planks
James Brough & Co.	Labour
James Brough & Co.	Labour
James Rabone	Lime & carriage
James Riddings & Co.	Slating
James Brough & Co.	Labour
James Rabone	5 strikes of hair
John Nixon	1,500 bricks
John Cork	Wheeling bricks
John Heywood, Joyclyn & Co.	Labour
John Grigg	Iron work
John Granger	For sundries
John Payne & Co.	Labour
John Iddins	Timber
John Payne & Co.	Labour
John Percival & Co.	Labour
John Cope	Timber
John Percival & Co.	Labour
John Wilkinson	Cast iron goods
John Nixon	Hauling a boat of bricks
John Grigg	Ironwork
Jos Blakemoor	Carriage of cylinder
Joseph Astbury	Slabs
Joseph Warr	Loading bricks
Offten & Co.	Labour
Owen	Carriage
Richard Turley	Two loads of lime
Richard Russel	For wheelbarrows
Rick Turvey	Lime
Sam Watson	For a rope
Samuel Rudge & Co.	Labour
Samuel Rudge & Co.	Working the new engine
Southall & Co.	Bellows
Thomas Salmon	Locks, bolts etc.
Walter Adams	Carriage
Westley Masdon	Plumming & lead
William & Co.	Labour
William & Stephen Hipkiss	A boiler

Fig. 116: The list of companies involved in the construction of the Smethwick Engine and associated civil engineering works produced by Mr J H Andrews of the Birmingham Museum of Science and Industry.

Engine & pump components	Weight in		
	Cwt	Qtr	Lb
4 engine bars	1	2	3
Nozel	12	3	5
Upright steam pipe	5	1	24
Four 26" pipes	76	2	0
One clack door pipe	27	0	8
One 9" pipe 8'4" long	5	2	26
One 9" pipe elbowed	2	3	5
One 10" pipe 10ft long	8	3	0
One clack door plate	4	2	25
One clack	4	0	6
Sub-total	149	1	18
@ 18/- per Cwt = £134-17s-10d			
One gudgeon plate	4	2	0
Two plummer blocks	5	0	4
Sub-total	9	2	4
@ 16/- per Cwt = £8-11s-4d			
One 33" cylinder	36	3	14
Top of cylinder turned	5	3	4
Piston to cylinder turned & coned	4	3	18
Stuffing box	0	2	3
Inner bottom turned	5	3	0
Outer bottom	5	2	22
Eduction or steam pipe foot	1	1	20
Air pump & bucket	6	3	10
Round cover for air pump	0	3	25
Top box	2	3	25
Water pump	6	2	0
Bucket for water pump & door	0	3	26
Little lid	0	1	22
Air pump bottom	1	2	26
One bucket 24" Dia.	2	1	13
Sub-total	83	3	4
@ 30/- per Cwt = £125-13s-6d			
Two brass weights for stuffing box	0	1	0½
Two brass valves for air pump	0	0	5½
Brass valve in top box	0	0	5½
Brass stuffing box for air pump	0	0	23½
Sub-total	0	2	7
@ 1/- per lb = £3-3s-0d			

Fig. 117: The company which supplied the castings and almost all of the mechanical parts for the Smethwick Engine was John Wilkinson's Bersham Ironworks. This is his entry reproduced in full.

finished it on Thursday 27th May 1779 when the engine was steamed for the first time, and without any of the modifications which the engine underwent during its long working life.

The Cost of the Construction of the Smethwick Engine 1778-1779

The one document which proved of inestimable value when trying to discover how the Smethwick engine would have looked in May 1779. Fig. 116 shows a complete list of the companies involved in the construction of the Smethwick Engine. Each of these companies invoiced the canal company for parts supplied or work done on the site during 1778-1779. Fig. 117 is a detailed account of the major supplier, John Wilkinson, who supplied the main cylinder and many of the engine parts from his Bersham Ironworks. These lists were compiled by Mr J H Andrews of the Birmingham Museum of Science and Industry.

Mr Andrews studied the accounts of the Birmingham Canal Navigation Company and he was able to produce a complete record of the cost of the 32" diameter engine erected at Smethwick during 1778 and 1779. The accounts were extracted from the journals of payments and the cash books of the canal company which are now deposited in the Public Record Office at Kew. Not only did these records give the costs of the original engine, but in many of the entries a good description and the actual sizes of many of the items are given. Fifty-six separate companies were involved with the construction of the engine and the complementary civil engineering work needed to replenish the canal system with water.

Many of the companies working for the Birmingham Canal Navigation Company on the installation of this pumping engine performed a great diversification of tasks, for example Samuel Rudge & Co. invoiced the Canal Company on forty-seven occasions and, although this was mainly for the labour involved, the records did show a price of £2-2s-11d to work the new engine. The total amount invoiced by Samuel Rudge & Co. for the whole installation was £493-6s-6½d.

It is not known how many men were involved in the construction of the original engine and engine house, but at least one woman was involved, because Hannah Hipkiss charged the canal company £4-2s-0d for hauling loads of bricks and iron castings by boat. Many of the heavy items were delivered to the erection site by canal boat. One entry states that James Rabone supplied 170,000 bricks for a cost of £92-3s-8d. The company which supplied the castings, and almost all of the mechanical parts, was John Wilkinson from his Bersham Ironworks. This entry is reproduced in full because it reveals many of the component sizes to the original engine.

The final payment was to the company who started this project – Matthew Boulton and James Watt. They invoiced the canal company for £210 which was to be paid annually, and the amount was calculated from the saving in fuel which would have been used if a standard atmospheric engine had been installed, without a separate water cooled condenser.

Although the last entries in the cash book had a date of 18 September 1779, James Watt undertook extensive trials on the engine in June 1779. The miniature version of the Smethwick Engine took *much* more time to construct than the original engine as every item had to be hand-made. Almost 3½ years, or 4,200 hours, were spent on the research, planning and construction of the model.

The first payment for the construction of the Smethwick installation was made to Samuel Rudge on 1st August 1778 for tunnelling. On the 5th June 1779 Robert and James Barber were paid £2-5s-0d for one boat load of lump coal for working the new engine. This is said to have been a load of 22 tons.

Conclusion to the Smethwick Engine in 1779

The final costs were:

Total payments for the engine............................ £1089-16s-4d
Total payments for the engine house £450-18s-7d
Total payments for tunnels, shafts and feeder £554-19s-3d

Total costs for the engine installation **£2095-14s-2d**

BIBLIOGRAPHY

A Treatise on the Steam Engine by John Farey, Volume 1 (1827, reprinted David and Charles 1971)

James Watt and the Steam Engine. H. W. Dickinson & R. Jenkins (1927)

The Steam Engine of Thomas Newcomen, L. T. C. Rolt. & J. A. Allen (1977)

Rees's Manufacturing Industry (1819-20), Volume Five

A Course of Experimental Philosophy, J. T. Desaguliers (1734-44)

A Short History of the Steam Engine, H. W. Dickinson (second edition, 1963)

Short Description of the Atmospheric Engine, M. Triewald, published in Stockholm, 1734

A History of Metallurgy, R. F. Tylecote, The Metal Society, 1976.

The Soho Foundry, Birmingham, W. K. V. Gale, W. & T. Avery Ltd. (published 1946)

Dictionary of Ceramics, Dodd, A. E., British Ceramic Research Association, George Newnes Ltd. (published 1964)

Transactions of the Newcomen Society between 1961 and 1998

A History of English Brickwork, Lloyd Nathaniel, Antique Collectors Club Ltd (reprinted 1983)

The Cornish Beam Engine, Barton, D. B., republished in 1989 by Cornwall Books

James Watt, L. T. C. Rolt, B. T. Batford Ltd. (published 1962)

GLOSSARY OF TERMS

Adze
A hand tool for cutting away the surface of wood like an axe with an arched blade at right angles to the handle.

Atmospheric Engine
The earth's atmospheric pressure is used to power an engine by pressing against a vacuum created on the underside of the piston.

Blister Steel
High carbon steel produced by the cementation process.

Bore
A cylindrical hole usually containing a sliding plug or piston.

Caulking
A means of closing the seams between wrought iron plates by the use of a blunt chisel.

Condensate
Water produced after condensation of the steam inside the cylinder or a separate condenser.

Condenser
A sealed vessel in which steam is condensed to create a vacuum.

Corves
Basket to put coal in, a man who made them was called a corver.

Double Acting
An engine with the piston powered in both directions.

Expansive Force
The force produced by steam expanding within a cylinder.

Factory System
A system which came into use in the eighteenth century for the manufacture of artefacts which were all produced on one site.

Firing Process
Reduction firing conditions created by burning coal to generate the heat for the firing process (reducing the oxygen).
Oxidizing firing conditions are created by generating the heat for the firing process by electric heating elements (oxygen present).

Flap Valve
A valve used for the one way passage of water and usually closed by the force of gravity.

Gib and Cotter
A metal attachment used to hold two components into a working position.

Granulate Pressing
A new method of forming pottery developed in the 1980s.

Great Lever
A term used to describe the main oscillating beam of a steam engine.

Hardware
Small ware or goods usually made from metal, eg. ironmongery.

Hogs Heads
A measure of a volume of water – 53 Imperial gallons.

Indicated Horsepower
Is calculated by the formula
$$P \times L \times A \times N / 33,000$$
P = mean pressure on the piston (p.s.i.).
L = length of the piston stroke in feet.
A = area of the piston in square inches.
N = number of strokes per minute.
33,000 = work done in foot pounds per minute and is equal to one horse power.

Injection Valve
A valve used to admit a controlled volume of water into the condenser.

GLOSSARY

Iron Cement
A compound of iron filings moistened with sal ammoniac which was used to seal the joints on cast iron pipes.

Marl
A brick clay found in the carboniferous system and used for the manufacture of bricks.

Mucksand
A material which is used to make cores used in sand casting - main ingredient horse manure.

Natural Philosophy
The forerunner of modern physics.

Normal Horsepower
Is an obsolescent term once used to rate the power of an engine.

Nozzle
A metal housing which is placed at each end of the powering cylinder containing the steam valves.

Oakum
A fibre obtained by the untwisting of old rope used for sealing joints to stop leaking.

Patent
A licence which grants the sole right to make and sell an invention.

Pickle Pot Condenser
A condenser used on the later Newcomen engines.

Piston
A sealed metal plunger which slides in the bore of a cylinder.

Plug Tree
A vertical rod attached to the main beam of an engine, used to operate the valve gear positioned below.

Plummer Block
A block of metal usually cast iron which is held into position by bolts and also containing a bearing.

Power of Engines
Long hundredweights Thomas Newcomen's method of calculating an engines power output by, using the pressure of the earths atmosphere. James Watt calculated the power of his engines by relating them to how much work a standard horse could do in one minute.

Preventer
An attachment bolted onto the main beam of an engine used to prevent damage in any emergency.

Primary Pump
A pump which is used to draw the water needed to run an engine from a well or stream.

P.S.I.
Pounds per square inch, usually used to calculate the force exerted by a piston.

Pug
A term used to describe a machine which extrudes clay.

Receiving Tank
A tank used to contain the water from the air pump.

Saggar
A refractory container used to hold unfired pottery for the firing process.

Secondary or Hot Water Pump
A pump which is used to raise the water from the receiving tank into the header tank of the engine.

Single Acting
An engine where the piston is powered in one direction only, usually by a vacuum on the underside.

Slip
A term used to describe liquid clay.

Snifting Valve
A valve designed to retain a vacuum, and allow the incoming steam to completely fill the cylinder.

Stuffing Box
A sealing box packed with hemp used to seal a circular sliding rod, usually the piston rod.

Tappets
Adjustable attachments on the side of the plug tree which were used to operate the levers controlling the flow of steam to the powering cylinder.

Vacuum
A space which has had the air removed, or at a very low pressure.

Vacuum Gauge
An instrument for checking the pressure inside the condenser, which is calibrated in inches of mercury.

INDEX

INDEX

157